ENVIRONMENTAL RESEARCH ADVANCES

ENVIRONMENTAL LITIGATION AND THE EPA

TRENDS, COSTS, ISSUES

ENVIRONMENTAL RESEARCH ADVANCES

Additional books in this series can be found on Nova's website under the Series tab.

Additional e-books in this series can be found on Nova's website under the e-book tab.

ENVIRONMENTAL RESEARCH ADVANCES

ENVIRONMENTAL LITIGATION AND THE EPA

TRENDS, COSTS, ISSUES

MARTIN F. OLIVER
EDITOR

nova publishers
New York

Library of Congress Cataloging-in-Publication Data

ISBN: 978-1-63482-720-1

Published by Nova Science Publishers, Inc. † New York

CONTENTS

PREFACE

In 2007, the Supreme Court handed down *Massachusetts v. EPA*, its first pronouncement on climate change and arguably the most important environmental law decision of the past decade. This book reviews that decision and examines trends in and factors affecting environmental litigation for fiscal years 1995 through 2010. It also reviews the Department of Justice's recent costs and recent plaintiff payments from the Judgment Fund and Environmental Protection Agency.

In: Environmental Litigation and the EPA ISBN: 978-1-63482-720-1
Editor: Martin F. Oliver

Chapter 1

THE SUPREME COURT'S FIRST CLIMATE CHANGE DECISION: *MASSACHUSETTS V. EPA*[*]

Robert Meltz

SUMMARY

In 2007, the Supreme Court handed down *Massachusetts v. EPA*, its first pronouncement on climate change and arguably the most important environmental law decision of the past decade. This report reviews that decision, but leaves coverage of the many EPA actions based on the decision to other CRS reports.

Massachusetts v. EPA was a case brought to challenge EPA's denial of a petition asking the agency to regulate greenhouse gas (GHG) emissions from new motor vehicles under the Clean Air Act (CAA). By 5-4, the Court held first that Massachusetts had standing to sue, an issue that took up most of the majority opinion. On the merits, the Court found that the CAA definition of "air pollutant" was broad enough to include GHGs. That being so, the Court held, CAA Section 202 authorizes EPA to regulate emissions from new motor vehicles on the basis of their possible climate change impacts. Finally, the Court determined that the phrase "in his judgment" in Section 202 did *not* authorize EPA to inject

---

[*] This is an edited, reformatted and augmented version of a Congressional Research Service publication RS22665, prepared for Members and Committees of Congress dated March 10, 2014.

policy considerations into its decision whether to so regulate. For these reasons, the Court reversed the lower court decision upholding the petition denial.

The Court's decision left EPA with three options for responding to the petition: (1) find that new motor vehicle GHG emissions may "endanger public health or welfare," the prerequisite to limiting them under Section 202, then issue emission standards; (2) find that they do *not* satisfy that prerequisite, or (3) decide that climate change science is so uncertain as to preclude making either finding (1) or (2). Given the state of climate change science by 2007, it was widely believed at the time that option (1) was the only legally defensible one for EPA. This is the option that EPA took, as covered in other CRS reports.

A few legal comments may be made about the *Massachusetts v. EPA* decision from the vantage point of 2013. At the outset, the finding of standing in Massachusetts has not proved helpful to non-state plaintiffs in climate change litigation, leaving intact this considerable threshold hurdle for climate change plaintiffs. In addition, the *Massachusetts* holding was used by a 2011 Supreme Court decision to bar federal common law claims against entities based on their contribution to climate change. On the other hand, *Massachusetts* helped bring about a 2010 litigation settlement that committed EPA to restricting GHG emissions from certain *stationary* sources of emissions, which it has now set about doing.

Now six years old, the *Massachusetts* decision remains judicially unquestioned. Its holding that the CAA authorizes EPA to regulate GHG emissions remains the governing law, barring Supreme Court reversal or congressional amendment of the CAA. Quite recently, the Supreme Court agreed to hear its third climate change case, but limited argument to a narrow issue, making it very unlikely the Court will disturb *Massachusetts*.

In 2007, the Supreme Court handed down *Massachusetts v. EPA*, its first pronouncement on climate change and arguably its most important environmental law decision in the past decade.[1] By a narrow 5-4 margin, the Court held that the Environmental Protection Agency (EPA) in 2003 had improperly denied a petition asking the agency to regulate greenhouse gas (GHG) emissions from new motor vehicles under the Clean Air Act (CAA). Contrary to EPA's position, the Court said that the CAA definition of "air pollutant" was broad enough to include GHGs. Accordingly, the Court reversed the lower court decision upholding the petition denial.

The Supreme Court decision did not *compel* EPA to regulate greenhouse gas (GHG) emissions from new motor vehicles, but it did limit the range of options available to the agency so that doing so was its most defensible course of action.

This report confines itself to the *Massachusetts v. EPA* litigation and leaves to other CRS reports the numerous EPA actions taken as a result of the Supreme Court decision.[2] The report traces the events leading up to the Court's decision, describes the decision itself, notes some general implications, and then comments on the decision's continuing force.

EPA'S DENIAL OF THE SECTION 202 PETITION

The saga of *Massachusetts v. EPA* began in 1999. In that year, 19 environmental and energy organizations petitioned EPA to regulate emissions of GHGs (CO_2, methane, nitrous oxide, and hydrofluorocarbons) from new motor vehicles.[3] The petition cited the agency's alleged mandatory duty to do so under CAA Section 202(a)(1),[4] which directs the EPA Administrator to prescribe "standards applicable to the emission of any air pollutant from any class or classes of new motor vehicles ... which, in his judgment cause, or contribute to, air pollution which may reasonably be anticipated to endanger public health or welfare."

In 2003, after receiving about 50,000 comments, EPA denied this petition.[5] Much of the agency's rationale followed a General Counsel memorandum issued the same day.[6] Contrary to its Clinton Administration precursor,[7] this new General Counsel memorandum concluded that the CAA does *not* grant EPA authority to regulate CO_2 and other GHG emissions based on their climate change impacts. Thus the agency had no choice but to reject the petition, though it also had policy reasons for doing so.

MASSACHUSETTS V. EPA IN THE D.C. CIRCUIT

EPA's denial of the Section 202 petition prompted a suit, *Massachusetts v. EPA*, in the D.C. Circuit seeking review of the denial. Petitioners were 12 states (CA, CT, IL, MA, ME, NJ, NM, NY, OR, RI, VT, WA); three cities (New York, Baltimore, and Washington, D.C.); two U.S. territories (American Samoa and Northern Mariana Islands); and several environmental groups. Opposing the challenge, besides EPA, were 10 state intervenors (AK, ID, KS,

MI, ND, NE, OH, SD, TX, UT), plus several automobile- and truck-related trade groups.

In 2005, a split panel of the D.C. Circuit rejected the suit—in effect upholding EPA's denial of the petition.[8] The two judges supporting rejection of the suit, however, did so for different reasons. Judge Randolph concluded that EPA had properly exercised its discretion in choosing not to wield its Section 202 authority. As to this discretion issue, recall that CAA Section 202(a)(1) directs the EPA Administrator to prescribe standards for any motor vehicle emissions that "in his judgment" cause harmful air pollution. Judge Randolph read "in his judgment" broadly to allow EPA consideration of not only scientific uncertainty about the effects of GHGs but also policy considerations that justify not regulating. Thus EPA, in his view, was entitled to rely, as it did, on such factors as the George W. Bush Administration's policy preference for voluntary GHG control measures, and its belief that regulating motor vehicle emissions was a piecemeal, hence inefficient, approach to dealing with climate change. By contrast, Judge Sentelle, the other judge supporting rejection of the petition, simply held that petitioners lacked standing.[9]

In dissent, Judge Tatel asserted that Massachusetts had demonstrated standing through past and future loss of shore land as a result of climate-change-induced sea level rise. On the merits, he found that EPA has authority under Section 202(a)(1) to regulate GHG emissions. He further concluded that EPA's 202(a)(1) discretion does not extend to policy considerations, as Judge Randolph held, but relates exclusively to whether the emissions cause harmful air pollution.

Massachusetts v. EPA in the Supreme Court

It was somewhat of a surprise that the Supreme Court agreed to review the D.C. Circuit decision in *Massachusetts v. EPA*. There was no split in the circuits, which often disposes the Court not to take a case, and the D.C. Circuit majority had not even ruled on the key issue: whether CAA Section 202(a)(1) authorizes regulation of GHG emissions. Moreover, grants of certiorari over the opposition of the United States, as here, are rare. But as the Supreme Court stated in its decision, "the unusual importance of the underlying issue persuaded us to grant the writ."[10]

The Court ruled 5-4 for petitioner states and environmental groups on all three issues in the case. It held first that at least one petitioner, the Commonwealth of Massachusetts, had standing to sue, so the Court could proceed to the merits. On the merits, it found that the CAA gives EPA authority to regulate GHG emissions from new motor vehicles (and, by implication, from other emission sources), and does not give EPA discretion to inject policy considerations into its decision whether to so regulate. Justice Kennedy provided the fifth vote by joining Justice Stevens's opinion for the Court's "liberal/moderate" bloc. The ruling in favor of petitioners was forecast early in the majority opinion by its opening sentences: "A well-documented rise in global temperatures has coincided with a significant increase in the concentration of carbon dioxide in the atmosphere. Respected scientists believe the two trends are related."[11] Nor did the dissenters dispute this.

Standing to Sue

Most of the Supreme Court's decision is devoted to whether plaintiffs had standing to sue, an issue that is recurring in climate change litigation. At the outset, the Court found that petitioners had two factors in their favor. First, the CAA specifically authorizes challenges to agency action unlawfully withheld, such as the *Massachusetts* suit.[12] A litigant to whom Congress has accorded such a procedural right, said the Court, can assert that right without meeting the normal standards for standing.[13] Second, the Court found it "of considerable relevance" that the petitioner injury on which it focused—Massachusetts's loss of shore land from global-warming-induced sea level rise—was that of a sovereign state rather than a private entity.[14] States are "not normal litigants for the purposes of invoking federal jurisdiction," said the Court, noting their quasi-sovereign duty to preserve their territory.[15]

Having described petitioners' favored position with regard to standing, it was curious that the Court then undertook a fairly traditional standing analysis. As to the first prong of the black-letter standing test—whether plaintiff has demonstrated actual or imminent "injury in fact" of a concrete and particularized nature—the Court homed in on Massachusetts's status as owner of much of the commonwealth's shore land. That this injury may be widely shared with other coastal states does not disqualify this injury, said the Court; it is nonetheless concrete.[16]

The second prong of the standing test is causation, requiring that the injury of which the plaintiff complains is fairly traceable to the defendant. EPA did not dispute the existence of a causal relationship between GHG emissions and climate change. It did argue, however, that any reduction in GHG emissions achieved through the current litigation would be too tiny a fraction of worldwide GHG emissions to make a cognizable difference in climate change.[17] In an important ruling that may be of benefit to environmental plaintiffs in many contexts, the Court held that even an agency's refusal to take a "small incremental step" that would result in only a modest reduction in worldwide GHG emissions, is enough for standing purposes.[18]

The third and final prong of the standing test is redressability, demanding that the remedy sought by the plaintiff is one likely to redress his injury. In this case, the remedy sought was EPA regulation of GHG emissions from new motor vehicles. The Court found that this remedy satisfied redressability because while it would not by itself reverse climate change, it would nonetheless slow or reduce it.[19] Nor, given the "enormity" of the potential effects of climate change, was it relevant to the Court that the full effectiveness of the remedy would be delayed until existing cars and trucks on the road were largely replaced by new ones.[20]

The Clean Air Act Issues

Given the large number of pages devoted by the majority opinion to standing, its discussion of the two CAA issues in the case seems strangely brief.

On the question of EPA's authority to regulate GHG emissions, the Court looked to the CAA's "sweeping" definition of "air pollutant," embracing "*any* air pollutant ... including *any* physical, chemical ... substance or matter which is emitted into or otherwise enters the ambient air."[21] Such a broad definition, it said, simply could not be squared with EPA's position that GHGs are not included.[22] The Court rejected EPA's argument that federal laws enacted following enactment of this statutory definition—laws emphasizing interagency collaboration and research—suggest that Congress meant to curtail EPA's power to use mandatory regulations in addressing air pollutants.[23] Nor was the Court moved by EPA's contention that "air pollutant" in the CAA could not include vehicle GHG emissions because EPA standards for such emissions could be satisfied only by improving fuel economy, a job EPA asserted was assigned solely to the Department of Transportation under a different statute (the Energy Policy and Conservation Act).[24]

As to the other, discretion issue, the Court concluded that the phrase "in his judgment" in CAA Section 202 should be read narrowly.[25] That is, it allows the EPA Administrator, in deciding whether to set emission standards, to consider only whether an air pollutant, in the section's words, "may reasonably be anticipated to endanger public health or welfare."[26] The phrase does not give EPA discretion to factor in its policy preferences. Policy considerations, at least those that led EPA to reject the petition, "have nothing to do with whether greenhouse gas emissions contribute to climate change."[27] Thus, said the Court, EPA can avoid taking further action in response to the Section 202 petition "only if it determines that greenhouse gases do not contribute to climate change or if it provides some reasonable explanation as to why it cannot or will not exercise its discretion."[28] In sum, whether EPA decides to act or not, "[it] must ground its reasons for action or inaction in the statute."[29]

Based on its resolution of the authority and discretion issues, the Court reversed the D.C. Circuit opinion, and remanded the case to that court. Months later, the D.C. Circuit vacated EPA's denial of the rulemaking petition and remanded the matter to the agency.[30]

A four-justice dissent by Chief Justice Roberts vigorously disputed the majority's finding of standing.[31] A four-justice dissent by Justice Scalia disputed that "air pollutant" in Section 202 includes GHGs.[32]

SINCE THE SUPREME COURT DECISION

Since the high court ruling in *Massachusetts v. EPA*, EPA has begun building a broad regulatory edifice for controlling GHG emissions, beginning with a Section 202 "endangerment" finding in 2009.[33] Since this finding, the agency has regulated GHG emissions from new motor vehicles.[34] Under a different section of the CAA, the agency likely will be regulating soon GHG emissions from certain stationary sources. It is interesting to note that because the *Massachusetts* decision was 5-4, a different vote by a single Justice would have meant that the CAA would have been found *not* to give EPA authority over GHG emissions, and this entire regulatory edifice probably would not exist or be under consideration today. As indicated earlier, these post-*Massachusetts* agency actions are described elsewhere; here, only a few select points are made.

Standing. So far, the finding of standing in *Massachusetts* has not proved helpful to *non-state* plaintiffs seeking to establish standing in other climate change litigation. It was conjectured soon after the *Massachusetts* decision that the Court's finding of standing might be contingent on the existence of the *state* petitioner there, plus the CAA's explicit provision allowing the filing of administrative petitions. And indeed, climate change cases since *Massachusetts* involving non-state plaintiffs and non-CAA causes of action (such as common law nuisance) have rejected extending its lenient standard for standing.[35]

Regulation of GHG emissions under CAA sections other than 202. The *Massachusetts* ruling upholding CAA coverage of Section 202 GHG emissions contributed to a 2010 litigation settlement that committed EPA to establishing new source performance standards for GHG emissions from fossil-fuel-fired power plants, under CAA Section 111.[36] *Massachusetts* may have further reverberations in EPA's clean air program, given its discussion of CAA terms found not only in Sections 202 and 111. Examples include "air pollutant," "in his judgment," and "may reasonably be anticipated to endanger public health and welfare."[37]

Displacement of federal common law. The holding of *Massachusetts* was used by a 2011 Supreme Court decision to bar federal common law claims (such as nuisance) against entities on the basis of their contribution to climate change.[38] *Massachusetts* strengthened the argument that Congress in the CAA intended to leave no room for courts to develop overlapping federal common law restricting GHG emissions, since it made clear that a congressional enactment, the CAA, was available for that same purpose. Ironically, this result meant that the victory for the "environmental" side in *Massachusetts v. EPA* contributed to the defeat for that side in the federal common law case. Though the 2011 ruling of the Court involved plaintiffs seeking a damages remedy, the ruling has been held to displace federal common law actions seeking injunctive relief as well.[39] The availability of *state* common law claims for reducing GHG emissions remains an open question.

Massachusetts remains good law. Now six years old, the *Massachusetts* ruling remains in full effect. Its holding that the CAA authorizes EPA to regulate GHG emissions from new motor vehicles is the governing law, barring Supreme Court reversal or congressional amendment unlikely in the near term.

Recently, no less than nine petitions for certiorari were filed in the Supreme Court seeking review of a D.C. Circuit opinion upholding EPA GHG rules that trace back to *Massachusetts*.[40] In the Supreme Court, the case is styled *Utility Air Regulatory Group v. Environmental Protection Agency (UARG)*.[41] Though several of these petitions raised issues that would have allowed the Court to reconsider *Massachusetts*, the Court declined the invitation. The Court did indeed grant certiorari in *UARG*, but limited to a narrow question that left *Massachusetts* intact—at least as applied to coverage of GHG emissions from new motor vehicles under CAA Section 202 (directly at issue in *Massachusetts*) and to new source performance standards under CAA Section 111[42] (conceded by petitioners' counsel at the *UARG* oral argument). The *UARG* petitioners did argue to the Court, however, that *Massachusetts* notwithstanding, GHG emissions are not within the term "air pollutant" as that term is used in the "prevention of significant deterioration" portion of the CAA.[43] A decision in *UARG* is expected by June, 2014.

End Notes

[1] 549 U.S. 497 (2007). Recent years have seen an explosion of litigation involving climate change, Massachusetts v. EPA being just one (though the most important) example. See CRS Report R42613, Climate Change and Existing Law: A Survey of Legal Issues Past, Present, and Future, by Robert Meltz.

[2] See, e.g., CRS Report R41103, Federal Agency Actions Following the Supreme Court's Climate Change Decision in Massachusetts v. EPA: A Chronology, by Robert Meltz.

[3] Petition for Rulemaking and Collateral Relief Seeking the Regulation of Greenhouse Gas Emissions from New Motor Vehicles Under Section 202 of the Clean Air Act, filed October 20, 1999.

[4] 42 U.S.C. § 7521(a)(1).

[5] EPA, Control of Emissions from New Highway Vehicles and Engines, 68 Fed. Reg. 52922 (September 8, 2003).

[6] Memorandum from Robert E. Fabricant, EPA General Counsel, to Marianne L. Horinko, EPA Acting Administrator, EPA's Authority to Impose Mandatory Controls to Address Global Climate Change Under the Clean Air Act (August 28, 2003).

[7] Memorandum from Jonathan Z. Cannon, EPA General Counsel, to Carol M. Browner, EPA Administrator, EPA's Authority to Regulate Pollutants Emitted by Electric Power Generation Sources (April 10, 1998).

[8] 415 F.3d 50 (D.C. Cir. 2005).

[9] The test for whether a plaintiff in federal court has standing is described in greater detail in the following discussion of the Supreme Court's decision on appeal of the D.C. Circuit ruling.

[10] 549 U.S. at 506.

[11] Id. at 504-505.

[12] CAA § 307(b)(1), 42 U.S.C. § 7607(b)(1).

[13] 549 U.S. at 517-518.

[14] 549 U.S. at 518.

[15] Id.

[16] Id. at 522.

[17] Id. at 523.

[18] Id. at 524.

[19] Id. at 525.

[20] Id. at 525.

[21] Id. at 528-529 (emphasis added by Court). The CAA definition of "air pollutant" is in Section 302(g) of the act, 42 U.S.C. § 7602(g).

[22] 549 U.S. at 528.
[23] Id. at 529-530.
[24] Id. at 531-532. The Energy Policy and Conservation Act provision on which EPA relied is at 49 U.S.C. § 32902.
[25] 549 U.S. at 532-534.
[26] Id. at 532-533.
[27] 549 U.S. at 533.
[28] Id.
[29] Id. at 535.
[30] 249 Fed. Appx. 829 (D.C. Cir. 2007).
[31] 549 U.S. at 535-549.
[32] Id. at 555-560.
[33] 74 Fed. Reg. 66496 (2009).
[34] See, e.g., 75 Fed. Reg. 25323 (GHG emission standards for 2012-2016 model year light-duty vehicles); 76 Fed. Reg. 57106 (GHG emission standards for 2014 and later model year medium- and heavy-duty vehicles).
[35] See, e.g., Comer v. Murphy Oil USA, Inc., 839 F. Supp. 2d 849 (S.D. Miss. 2012), affirmed on other grounds, 718 F.3d 460 (5th Cir. 2013).
[36] 42 U.S.C. § 7411. The settled case is New York v. EPA, No. 06-1322 (D.C. Cir. September 13, 2006) (severed from pre-existing case by order of the court).
[37] See, e.g., CAA § 108(a)(1)-(2), 42 U.S.C. § 7408(a)(1)-(2) (requiring the EPA Administrator to maintain a list of each "air pollutant" "emissions of which, in his judgment, cause or contribute to air pollution which may reasonably be anticipated to endanger public health or welfare," and then issue air quality criteria and national ambient air quality standards for such pollutants).
[38] American Elec. Power, Inc. v. Connecticut, 131 S. Ct. 2527 (2011).
[39] Comer, 839 F. Supp. 2d 849.
[40] Coalition for Responsible Regulation, Inc. v. EPA, 684 F.3d 102 (D.C. Cir. 2012), cert. granted, 134 S. Ct. 418 (2013).
[41] Docket no. 12-1146.
[42] 42 U.S.C. § 7411.
[43] CAA Title I, Part C; 42 U.S.C. §§ 7470-7492.

In: Environmental Litigation and the EPA
Editor: Martin F. Oliver
ISBN: 978-1-63482-720-1

Chapter 2

ENVIRONMENTAL LITIGATION: CASES AGAINST EPA AND ASSOCIATED COSTS OVER TIME*

United States Government Accountability Office

WHY GAO DID THIS STUDY

The Environmental Protection Agency(EPA) faces numerous legal challenges as it implements the nation's environmental laws. Several statutes, such as the Clean Air and Clean Water Acts, allow citizens to file suit against EPA to challenge certain agency actions. Where EPA is named as a defendant, the Department of Justice provides EPA's legal defense. If successful, plaintiffs may be paid for certain attorney fees and costs. Payments are made from the Department of the Treasury's Judgment Fund—a permanent fund available to pay judgments against the government, as well as settlements resulting from lawsuits—or EPA's appropriations. For this review, GAO was asked to examine (1) the trends in and factors affecting environmental litigation for fiscal years 1995 through2010 and (2) Justice's recent costs and recent plaintiff payments from the Judgment Fund and EPA.

To conduct this review, GAO obtained and analyzed data from two Justice databases on cases filed under 10 key environmental statutes. To gain

* This is an edited, reformatted and augmented version of a United States Government Accountability Office publication, No. GAO-11-650, dated August 2011.

stakeholder views on any trends and factors that might affect them, GAO interviewed representatives of environmental and industry groups, state attorneys general, and other experts. GAO estimated the costs of litigation handled by Justice attorneys and payments made for attorney fees and court costs from the Judgment Fund and EPA funds.

GAO is making no recommendations in this report. GAO provided a draft of this report to the agencies for comment. Justice and Treasury had technical comments, which were incorporated, while EPA had no comments.

WHAT GAO FOUND

No trend was discernible in the number of environmental cases brought against EPA from fiscal year 1995 through fiscal year 2010, as the number of cases filed in federal court varied over time. Justice staff defended EPA on an average of about 155 such cases each year, or a total of about 2,500 cases between fiscal years 1995 and 2010. Most cases were filed under the Clean Air Act (59 percent of cases) and the Clean Water Act (20 percent of cases). According to stakeholders GAO interviewed, a number of factors—particularly a change in presidential administration, new regulations or amendments to laws, or EPA's not meeting statutorily required deadlines—affect environmental litigation.

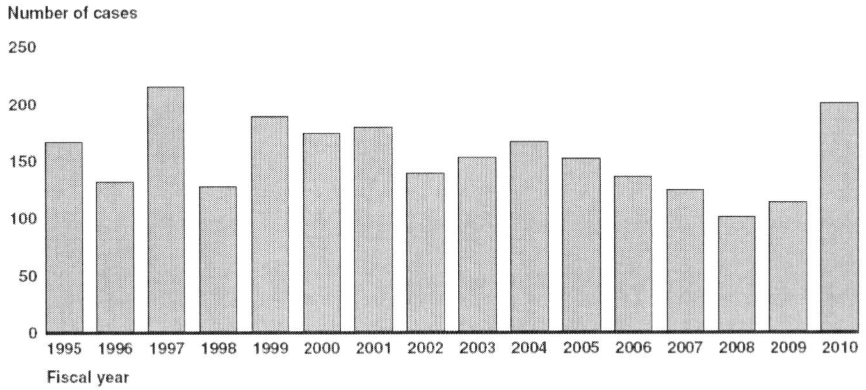

Source: GAO analysis of Justice data.

Environmental Litigation Cases Filed against EPA, Fiscal Year 1995 through Fiscal Year 2010.

The costs borne by Justice, EPA, and Treasury also varied without a discernible trend from fiscal year 1998 through fiscal year 2010. Justice spent at least $43 million, or $3.3 million annually, to defend EPA in court during this time. In addition, owing to statutory requirements to pay certain successful plaintiffs for attorney fees and costs, Treasury paid about $14.2 million from fiscal year 2003 through fiscal year 2010—about $1.8 million per fiscal year—to plaintiffs in environmental cases. EPA paid approximately $1.4 million from fiscal year 2006 through fiscal year 2010—about $280,000 per fiscal year—to plaintiffs for environmental litigation claims under relevant statutes. (All amounts are given in constant 2010 dollars.) Justice officials said that they negotiate payments with the successful plaintiffs, who generally receive less than originally requested. Complicating efforts to analyze trends in cases and costs is that Justice maintains data on environmental cases in two separate data systems and does not have a standard approach for maintaining the data. As a result, it is difficult to identify and summarize the full set of cases and costs managed by Justice. Nonetheless, using an iterative electronic and manual process, GAO was able to merge the two sets of data for its purposes. Justice officials said that they do not need to change their approach to managing the data, however, because they do not use it to summarize case data agencywide. Moreover, the officials said they lack resources to adapt their aging systems to accept additional data.

ABBREVIATIONS

EAJA	Equal Access to Justice Act
EPA	Environmental Protection Agency
NEPA	National Environmental Policy Act
PACER	Public Access to Court Electronic Records

* * *

August 1, 2011

The Honorable James M. Inhofe
Ranking Member
Committee on Environment and Public Works
United States Senate

The Honorable David Vitter
United States Senate

As the primary federal agency charged with implementing many of the nation's environmental laws, the Environmental Protection Agency (EPA) often faces the prospect of litigation over its regulations and other actions. For example, several environmental statutes have provisions that allow citizens—including individuals, states, companies, and associations—to file suit against EPA challenging certain agency actions, such as making regulations or permitting decisions. Where EPA is named as the defendant in lawsuits, the Department of Justice, which is generally responsible for defending federal agencies in court, provides EPA's legal defense, and EPA provides technical expertise. Within Justice, the Environment and Natural Resources Division handles most of the defense work on EPA environmental litigation cases from its Washington, D.C., office, but some of the 94 U.S. Attorneys' Offices, particularly those in the New York City area, also handle a small number of cases and may work on some cases managed by the Environment and Natural Resources Division.

Statutes establishing programs administered by EPA, and under which the agency may be sued, include many of the nation's most prominent environmental laws, such as the Clean Air Act; Clean Water Act; Comprehensive Environmental Response, Compensation, and Liability Act (better known as the Superfund law); Emergency Planning and Community Right-to-Know Act; Federal Insecticide, Fungicide, and Rodenticide Act and related provisions of the Federal Food, Drug, and Cosmetic Act; Resource Conservation and Recovery Act; Safe Drinking Water Act; and Toxic Substances Control Act.[1] In addition, EPA may be sued to challenge the agency's compliance with section 7 of the Endangered Species Act, which requires every federal agency, including EPA, to consult with the federal agencies responsible for species management if it is proposing an action—including a major construction project or a regulation or permit—that may affect species protected under the act or habitat designated as critical to the species' survival.

In general, plaintiffs and defendants in lawsuits bear their own litigation costs, although some statutes authorize judges to award attorney fees to a successful, or prevailing, plaintiff. The same holds true in cases involving the federal government: if plaintiffs win a lawsuit against a federal agency, they cannot be awarded attorney fees and court expenses unless authorized by law. Both Justice and EPA incur costs in association with litigation, as does the plaintiff. The types of costs include attorney fees and expenses and court costs. Attorney fees include costs for attorney representation, which is typically charged according to time spent on a case. Other attorney expenses may include costs for expert witnesses, telephone, postage, travel, copying, and computer research expenses. Court costs include fees charged by courts, such as filing fees and reporting fees. In addition, EPA and the plaintiff typically incur costs stemming from time spent on a case, such as meeting with attorneys, negotiations, preparation for trial, and other activities. Payment for authorized attorney fees and costs generally comes from the Department of the Treasury's Judgment Fund, which is a permanent, indefinite appropriation available to pay many judgments against, and settlements by, the United States, or from an agency's appropriated funds.[2]

Currently, no aggregated data on such environmental litigation or associated costs are reported by federal agencies. The key agencies involved—Justice, EPA, and the Treasury—maintain certain data on individual cases in several internal agency databases, but collectively, these data do not capture all costs. Each of Justice's litigation components maintains a separate case management system that gathers information related to individual cases, and we have previously reported that the department's decentralized data management systems make it difficult and costly to gather data across divisions.[3] Both Justice and EPA record some case-related payments, including certain data on attorney fees and court costs associated with cases, although neither agency currently reports these data publicly.[4] In addition, Treasury records data on payments made from its Judgment Fund but does not publish them.

In this context, you asked us to review environmental litigation data. Our objectives were to examine (1) trends, if any, in environmental lawsuits against EPA from fiscal year 1995 through fiscal year 2010, as well as stakeholders' views of factors affecting any trends, and (2) Justice's recent costs for representing EPA in defensive environmental lawsuits and the federal government's recent payments to plaintiffs.

To conduct our work, we obtained and analyzed historical data from two components within Justice—Justice's Environment and Natural Resources Division, which defends EPA in most environmental litigation cases, and the Executive Office for U.S. Attorneys, which manages data for the 94 U.S. Attorneys' Offices. We gathered data from the Environment and Natural Resources Division's Case Management System database that tracks basic information on cases, including lead plaintiffs' names, filing and disposition dates, and relevant statutes. We also collected data from the Executive Office for U.S. Attorneys' Legal Information Office Network System, a database that contains similar—but not exactly the same—information for cases handled in part or in entirety by the 94 U.S. Attorneys' Offices. At the start of our engagement, Justice officials informed us that their databases were designed for management purposes rather than trend analysis. Nevertheless, since these databases were the best available sources of information on cases against EPA, we worked with Justice's Environment and Natural Resources Division and the Executive Office for U.S. Attorneys to use the data. From both databases, we examined data on lawsuits filed in federal court from October 1, 1994, through September 30, 2010 (i.e., fiscal year 1995 through fiscal year 2010) that were identified as brought under 10 environmental statutes and led by EPA. We excluded cases involving the Freedom of Information Act and contract, employment, or other generally applicable laws under which EPA may also have been sued. We also excluded cases filed under the National Environmental Policy Act (NEPA), which applies to major federal actions significantly affecting the environment, because, according to Justice, few NEPA cases are filed against EPA, and such cases are generally handled by the U.S. Attorneys' Offices and other sections of the Environment and Natural Resources Division, such that significantly more time would be required to include them.

To get their views on any environmental litigation trends and the factors that underlie them, we interviewed a nonprobability sample of the following stakeholders: officials from EPA and Justice; representatives of six environmental groups, six industry associations, and the National Association of Attorneys General; representatives of six state attorneys general or state environmental offices; and a university law professor who is expert in data on citizen suits. The findings from our interviews with stakeholders cannot be generalized to those to whom we did not speak.

We estimated the costs of environmental litigation on the basis of the following three factors: (1) Justice attorneys' labor for fiscal year 1998 through fiscal year 2010, (2) payments made by Treasury from the Judgment Fund for plaintiffs' attorney fees and court costs from fiscal year 2003 through fiscal year 2010, and (3) payments made from EPA appropriations from fiscal year 2006 through fiscal year 2010 (the years for which relevant payment data were available). Because EPA does not track its attorneys' time by case, we were not able to include data on EPA attorney costs spent on environmental litigation cases. In addition, the government may also incur other costs associated with litigation, including the costs of revising regulations in response to lawsuits, EPA overhead costs, and costs associated with delays in EPA permitting, but we did not have reliable data to quantify these costs.

For both Justice databases, we assessed the data and found them to be sufficiently complete and accurate for the purposes of this report. When certain case information was found missing or incorrect in some of the data fields, we did additional research on these cases using the federal courts' electronic records database and corrected the data. Inconsistent formatting of key data elements produced significant problems for completing our analysis and required significant manual review by GAO and Justice.

We conducted this performance audit from June 2010 through July 2011, in accordance with generally accepted government auditing standards. Those standards require that we plan and perform the audit to obtain sufficient, appropriate evidence to provide a reasonable basis for our findings and conclusions based on our audit objectives. We believe that the evidence obtained provides a reasonable basis for our findings and conclusions based on our audit objectives.

BACKGROUND

To carry out its responsibilities under the nation's environmental laws, EPA conducts an array of activities, such as promulgating regulations; issuing and denying permits; approving state programs; and issuing enforcement orders, plans, and other documents. Many of these activities may be subject to legal challenge.[5]

Environmental Statutes and Lawsuits against the Federal Government

Generally, the federal government has immunity from lawsuits, but federal laws authorize three types of suits related to EPA's implementation of environmental laws.[6] First, most of the major environmental statutes include "citizen suit" provisions authorizing citizens—including individuals, associations, businesses, and state and local governments—to sue EPA when the agency fails to perform an action mandated by law. These suits are often referred to as "agency-forcing" or "deadline" suits. Second, the major environmental statutes typically include judicial review provisions authorizing citizens to challenge certain EPA actions, such as promulgating regulations or issuing permits. Third, the Administrative Procedure Act[7] authorizes challenges to certain agency actions that are considered final actions, such as rulemakings and decisions on permit applications. As a result, even if a particular environmental statute does not authorize a challenge against EPA for a final decision or regulation, the Administrative Procedure Act may do so. Table 1 lists key environmental laws under which EPA takes actions—or that govern EPA actions—that may be subject to challenge in court.[8]

Table 1. Selected Environmental Laws

Statute	General topic	Examples of EPA actions subject to legal challenge
Safe Drinking Water Act	Public drinking water systems	Promulgation of national primary drinking water regulations Approval of state programs Administrative enforcement actions
Clean Water Act	Water pollution, dredging, and filling of waters	Promulgation of regulations Issuance and denial of permits Approval of state programs and certain actions by stateagencies Administrative enforcement actions
Clean Air Act	Emission of pollutants into air	Promulgation of national ambient air quality standards Promulgation of air quality regulations Issuance and denial of permits Approval of state programs and state implementation plans Administrative enforcement actions
Resource Conservation and Recovery Act	Solid and hazardous waste storage, handling, treatment, and disposal	Promulgation of regulations Issuance and denial of permits Approval of state programs Administrative enforcement actions
Comprehensive Environmental Response, Compensation, and Liability Act	Releases of hazardous substances and their cleanup	Cleanup remedies (after implementation) Promulgation of regulations
Toxic Substances Control Act	Toxic substances used in commerce	Promulgation of regulations Approval and denial of chemical registrations Administrative enforcement actions

Statute	General topic	Examples of EPA actions subject to legal challenge
Emergency Planning and Community Right-to-Know Act	Reporting releases of toxic chemicals	Promulgation of regulations Administrative enforcement actions
Federal Insecticide, Fungicide, and Rodenticide Act	Pesticides	Promulgation of regulations Approval and denial of pesticide applications
Federal Food, Drug, and Cosmetic Act	Pesticides used on food	Establishment of pesticide tolerances
Endangered Species Act	Agency actions affecting threatened and endangered species	Any actions taken under a substantive law may triggeraction under the Endangered Species Act

Source: GAO.

Supporters of provisions allowing legal challenges to actions of the federal government assert that they provide a check on the authority of federal agencies as they carry out—or fail to carry out—their duties. For example, in passing the 1977 Clean Air Act amendments, a key sponsor indicated that authorizing citizens to sue agencies to compel them to carry out their duties is integral to a democratic society.[9] According to others, citizen suits against government agencies have achieved benefits, such as ensuring the implementation of congressional directives or accelerating regulatory programs.[10] Similarly, the Administrative Procedure Act arose out of the expansion of the federal government in the New Deal, with concerns about agencies' adjudicative powers, their exercise of delegated legislative power by rulemakings, and the scope of review of agency administrative action by courts.

A lawsuit challenging EPA's failure to act may begin when the aggrieved party sends EPA a notice of intent to sue, if required, while a lawsuit challenging a final EPA action begins when a complaint is filed in court.[11]

Before EPA takes final action, the public or affected parties generally have opportunities to provide comments and information to the agency. In addition, administrative appeals procedures are available—and in many cases required[12]—to challenge EPA's final action without filing a lawsuit in a court.[13] For example, citizens can appeal an EPA air emission permit to the agency's Environmental Appeals Board. These administrative processes provide aggrieved parties with a forum that may be faster and less costly than a court.

If a party decides to pursue a case, the litigation process generally involves filing of a complaint, formal initiation of the litigation; motions to the court before trial, such as asking for dismissal of the case; and hearings and court decisions. Throughout this process, the parties to the litigation can decide to reach a settlement. Negotiations between the aggrieved party and EPA may occur anytime after the agency action, at any point during active litigation, and even after judgment. A common remedy sought in litigation against EPA under the statutes listed in table 1 is for the court to set aside an EPA regulation or permit decision and to require EPA to reconsider that regulation or permit decision.

The Equal Access to Justice Act

In the United States, parties involved in litigation generally pay their own attorney fees and costs, except in instances in which Congress has provided exceptions for policy reasons, such as to encourage citizens to bring suits to enforce the law. In these instances, as well as some common-law exceptions, a prevailing plaintiff may seek award of its attorney fees and court costs from the losing party. Many of the environmental statutes in table 1 contain such exceptions authorizing courts to award fees, which, according to Justice, include awards against the federal government.[14]

In 1980, Congress enacted the Equal Access to Justice Act (EAJA) authorizing the award of attorney fees and costs to parties that prevail in certain lawsuits against the federal government; the payments are made from Treasury's Judgment Fund and agency appropriations. While the federal government was already subject to some of these exceptions in environmental statutes, before EAJA was enacted, the federal government in many other cases was not subject to these exceptions and therefore was not authorized to make payments to prevailing parties. As the 1980 conference committee report for EAJA explains, the act's premise is that individuals, corporations, partnerships, and labor and other organizations do not seek review of or defend against unreasonable government actions because of the expense involved, as well as a disparity in expertise and resources between the government and the individual or organization involved.[15] For those cases brought under statutes that do not make the federal government subject to pay fees and costs, EAJA thus allows payment of the attorney fees and other costs if the organizations sought review of a government action and prevailed.[16] (See app. II for a detailed description of the act.)

Except as otherwise specifically provided by law, EAJA authorizes the award of the following costs to be paid from Treasury's Judgment Fund or an agency's appropriations, as indicated:

- *Court costs of prevailing parties* against the United States in any civil action. These costs may include fees for the clerk and marshal, reporter, printing, witnesses, copies, docket fees, and interpreters and court-appointed experts and may include an amount equal to the filing fees. Payment of costs made under this section generally are paid by Treasury's Judgment Fund.
- *Reasonable attorney fees and expenses of a prevailing party* to the same extent as any other party where a statutory or common-law exception provides for award of fees to a prevailing party.[17] Regarding the environmental statutes in table 1, according to Justice, many of the relevant provisions under which EPA may be sued provide for award of such fees against EPA, independent of EAJA. Nevertheless, EAJA makes EPA subject to fee awards under all the environmental statutes' provisions authorizing courts to award attorney fees and expenses. Therefore, in many—but not all—of the environmental lawsuits against EPA, a court may award attorney fees and expenses of a prevailing party against the agency, independently or as a result of EAJA section 2412(b).[18] Payment of awards made under this section generally are paid by Treasury's Judgment Fund.
- *Attorney fees and expenses of a prevailing party* in most other cases—that is, when the relevant statute does not authorize courts to award attorney fees and expenses, and no common-law exception applies—unless the court finds that the position of the United States was substantially justified or that special circumstances make an award unjust.[19] Two laws listed in table 1—the Federal Insecticide, Fungicide, and Rodenticide Act and the Federal Food, Drug, and Cosmetic Act—as well as some individual provisions of other statutes, do not authorize payment of fees to prevailing parties. As a result, in cases brought against EPA under these statutes and provisions, courts award payment of fees under EAJA section 2412(d).[20] Payment of awards made under this section is generally made from agency appropriations.

In addition, to settle a case, the government may agree to pay a plaintiff court costs and attorney fees and expenses. Payments made in connection with settlements are paid in the same manner as a court award for the case.

Some in Congress have expressed concerns that the use of taxpayer funds to make EAJA payments depletes limited funding; these individuals have called for transparency of these expenditures. Originally, EAJA provided for governmentwide reporting on its use and cost. For judicial proceedings, EAJA required the Director of the Administrative Office of the U. S. Courts to report annually to Congress on EAJA court activity, including the number, nature, and amounts of awards; claims involved; and any other relevant information deemed necessary to aid Congress in evaluating the scope and effect of awards under the act.

The responsibility for this reporting was transferred to the Attorney General in 1992. In addition, EAJA required the Chairman of the Administrative Conference of the United States to submit an EAJA report annually to Congress on administratively awarded fees and expenses. Then, in December 1995, the Federal Reports Elimination and Sunset Act of 1995 repealed the Attorney General's reporting requirement for fees and expenses awarded under EAJA and also discontinued reporting of governmentwide administrative awards of fees and costs under EAJA after fiscal year 1994.[21]

We have previously reported certain governmentwide EAJA data, as well as data focused on selected agencies. In 1995, we reported data on the number of cases and amounts of awarded plaintiff attorney fees exceeding $10,000 against nine federal agencies for cases closed during fiscal years 1993 and 1994.[22]

In 1998, we provided information on the history of EAJA, the extent to which one provision of the act was used governmentwide from 1982 to 1994, and the provision's use by the Department of Labor and other agencies.[23] The governmentwide data for fiscal year 1994 showed, among other things, that the Departments of Health and Human Services and of Veterans Affairs accounted for most EAJA payments in court proceedings, under the provision that applies when the substantive law does not authorize award of attorney fees and costs.[24]

The Number of Environmental Litigation Cases against EPA Showed No Discernible Trend over 16 Years, and Stakeholders Stated That Various Factors Affected Yearly Numbers

The number of environmental litigation cases brought against EPA each year from fiscal year 1995 through fiscal year 2010 varied but showed no discernible trend. According to the stakeholders we interviewed, a number of factors—particularly presidential administration, the passage of new regulations or amendments to laws, or EPA's failure to meet statutory deadlines—affect the number of environmental litigation cases each year and the type of plaintiffs who bring them.

No Trend Was Discernible in the Number of Cases Brought against EPA from Fiscal Year 1995 through Fiscal Year 2010

The number of environmental litigation cases brought against EPA each year from fiscal year 1995 through fiscal year 2010 varied but did not change systematically over time. The average number of new cases filed each year was 155, ranging from a low of 102 new cases filed in fiscal year 2008 to a high of 216 cases filed in fiscal year 1997 (see fig. 1). From fiscal year 1995 through fiscal year 2001, the average number of new cases was 170; from fiscal year 2002 through fiscal year 2010, the average number of new cases was 144, a difference of 26 fewer new cases on average. The average number of new cases in these periods varied from the long-term average of 155 cases by less than 10 percent. In all, Justice defended EPA in nearly 2,500 cases from fiscal year 1995 through fiscal year 2010.

The greatest number of cases was filed in fiscal year 1997, which, according to a Justice official, may be explained by the fact that EPA revised its national ambient air quality standards for ozone and particulate matter in 1997, which may have caused some groups to sue. In addition, according to the same official, in 1997 EPA implemented a "credible evidence" rule, which also was the subject of additional lawsuits.[25] The fewest cases against EPA (102) were filed in fiscal year 2008, and Justice officials were unable to pinpoint any specific reasons for the decline. In fiscal years 2009 and 2010, the caseload increased.

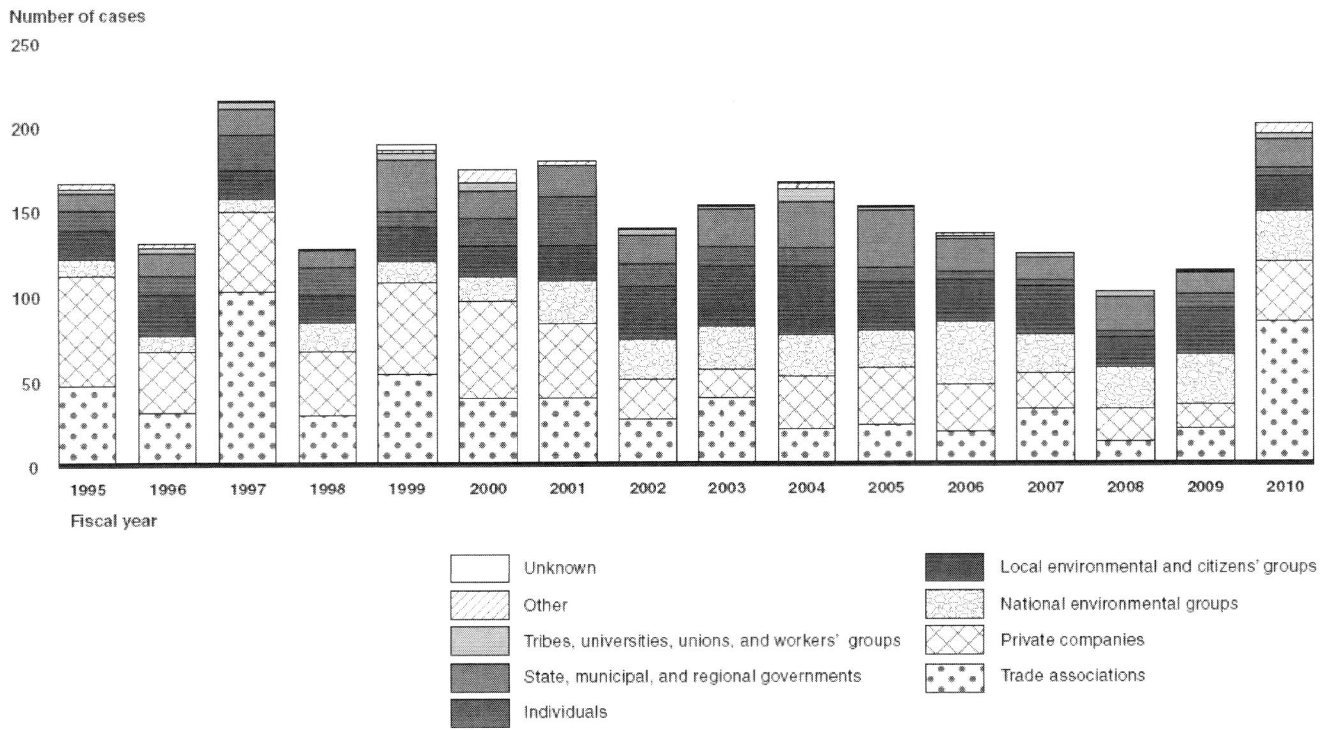

Source: GAO analysis of Justice data.

Figure 1. Environmental Cases Filed against EPA, Fiscal Year 1995 through Fiscal Year 2010.

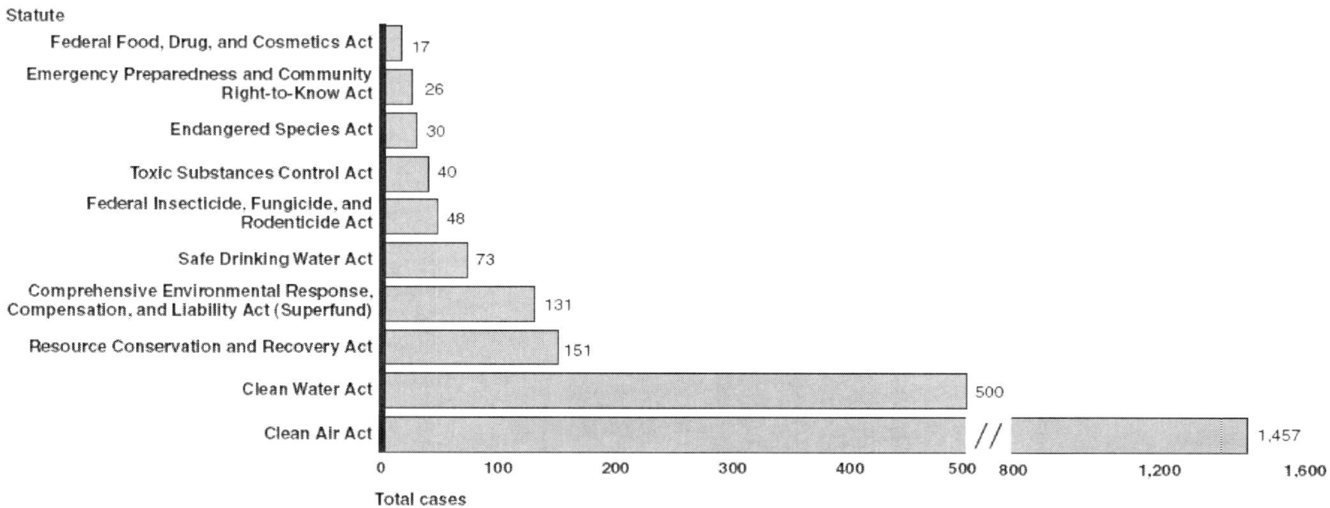

Source: GAO analysis of data from Justice's Environment and Natural Resources Division.
Note: Nine cases did not have information on statute.

Figure 2. Environmental Cases Filed against EPA by Statute, Fiscal Year 1995 through Fiscal Year 2010.

A Justice official said that it is difficult to know why the number of cases might increase because litigants sue for different reasons, and some time might elapse between an EPA action and a group's decision to sue.

As shown in figure 2, most cases against EPA were brought under the Clean Air Act, which represented about 59 percent of the approximately 2,500 cases that were filed during the 16-year period of our review. Cases filed under the Clean Water Act represented the next largest group of cases (20 percent), and the Resource Conservation and Recovery Act represented the third largest group of cases (6 percent).

The lead plaintiffs filing cases against EPA during the 16-year period fit into several categories. The largest category comprised trade associations (25 percent), followed by private companies (23 percent), local environmental groups and citizens' groups (16 percent), and national environmental groups (14 percent). Individuals, states and territories, municipal and regional government entities, unions and workers' groups, tribes, universities, and a small number of others we could not identify made up the remaining plaintiffs (see table 2). Appendix I gives more information about our method of developing these categories and classifying cases.

Table 2. Share of Cases by Lead Plaintiff Type: Fiscal Year 1995 through Fiscal Year 2010

Type of group[a]	Number of cases	Percentage
Trade associations	622	25
Private companies	566	23
Local environmental and citizens' groups	388	16
National environmental groups	338	14
States, territories, municipalities, and regional government entities	297	12
Individuals	185	7
Unions, workers' groups, universities, and tribes	46	2
Other	33	1
Unknown	7	1b
Total	2,482	100

Source: GAO.

[a]For more information on each of these groups, see appendix I. [b]Less than 1 percent.

Stakeholders Stated That Various Factors Influence Environmental Litigation

According to the stakeholders we interviewed, a number of factors—particularly a change in presidential administration, the passage of regulations or amendments to laws, and EPA's failure to meet statutory deadlines—affect plaintiffs' decisions to bring litigation against EPA. Stakeholders did not identify any single factor driving litigation, but instead, attributed litigation to a combination of different factors.

According to most of the stakeholders we spoke with, a presidential administration is an important factor in groups' decisions to bring suits against EPA. Some stakeholders suggested that a new administration viewed as favoring less enforcement could spur lawsuits from environmental groups in response, or industry groups could sue to delay or prevent the administration's actions. For example, a presidential administration that seems to favor less enforcement of requirements under environmental statutes could motivate increased litigation. Other stakeholders suggested that if an administration is viewed as favoring greater enforcement of rules, industry may respond to increased activity by bringing suit against EPA to delay or prevent the administration's actions, while certain environmental groups may bring suit with the aim of ensuring that required agency actions are completed during an administration they perceive as having views similar to the groups' own.

Most of the stakeholders also suggested that the development of new EPA regulations or the passage of amendments to environmental statutes may lead parties to file suit against the new regulations or against EPA's implementation of the amendments. When EPA issues new or amended regulations, parties may take issue with the specific new provisions. One stakeholder noted that an industry interested in a particular issue may become involved in litigation related to the development of regulations because it wishes to be part of the regulatory process and negotiations that result in a mutually acceptable rule. In addition, several of the stakeholders noted that if EPA does not meet its statutory deadlines, organizations or individuals might sue to enforce the deadline. In such suits, interested parties seek a court order or a settlement requiring EPA to implement its statutory responsibilities.

In addition, some stakeholders said that some statutes are broadly written or contain vague language or definitions; such statutes are more likely to be litigated because different parties want to define the terms and set precedent for future cases. For example, a stakeholder representing states' perspectives said that under the Clean Water Act, an area of frequent litigation is the definition of "navigable waters." Through lawsuits, litigants have argued about whether a certain body of water comes within the definition and can therefore be regulated under the act.

A few stakeholders identified two other factors that may affect litigation: (1) the maturity of the statute in question and (2) the use of existing laws to address new problems. The stakeholders said that the focus of litigation over a particular statute changes with time, as early cases may set precedents that will affect how the statute is implemented later. Also, a representative of an environmental organization said that because no major rewriting of any environmental statutes has occurred in 20 years, plaintiffs are increasingly bringing suits, and judges are making decisions, about how to interpret statutes in situations for which rules were not explicitly written. For example, parties disagree over whether the Clean Air Act should be used to regulate greenhouse gases, such as carbon dioxide, methane, and nitrous oxide—substances that some stakeholders say the act was not originally designed to regulate.

Available Data Indicate That Costs Associated with Environmental Litigation against EPA, including Payments to Plaintiffs, Have Varied over the Past 10 Years with No Discernible Trend

Data available from Justice, Treasury, and EPA show that the costs associated with environmental litigation cases against EPA have varied from year to year with no discernible trend. Justice's Environment and Natural Resources Division spent a total of about $43 million to defend EPA in these cases from fiscal year 1998 to fiscal year 2010, averaging $3.3 million per year.[26] Some cost data from the Department of Justice are not available, however, in part because Justice's Environment and Natural Resources Division and the U.S. Attorneys' Offices do not have a standard approach for maintaining key data for environmental litigation cases. For example, while the Environment and Natural Resources Division tracks attorney hours by case, the U.S. Attorneys' Offices do not. Treasury paid a total of about $14.2 million to prevailing plaintiffs for attorney fees and costs related to these cases from fiscal years 2003 through 2010, averaging about $1.8 million per year. EPA paid a total of $1.4 million from fiscal year 2006 through fiscal year 2010 in attorney fees and costs, averaging about $280,000 per year.

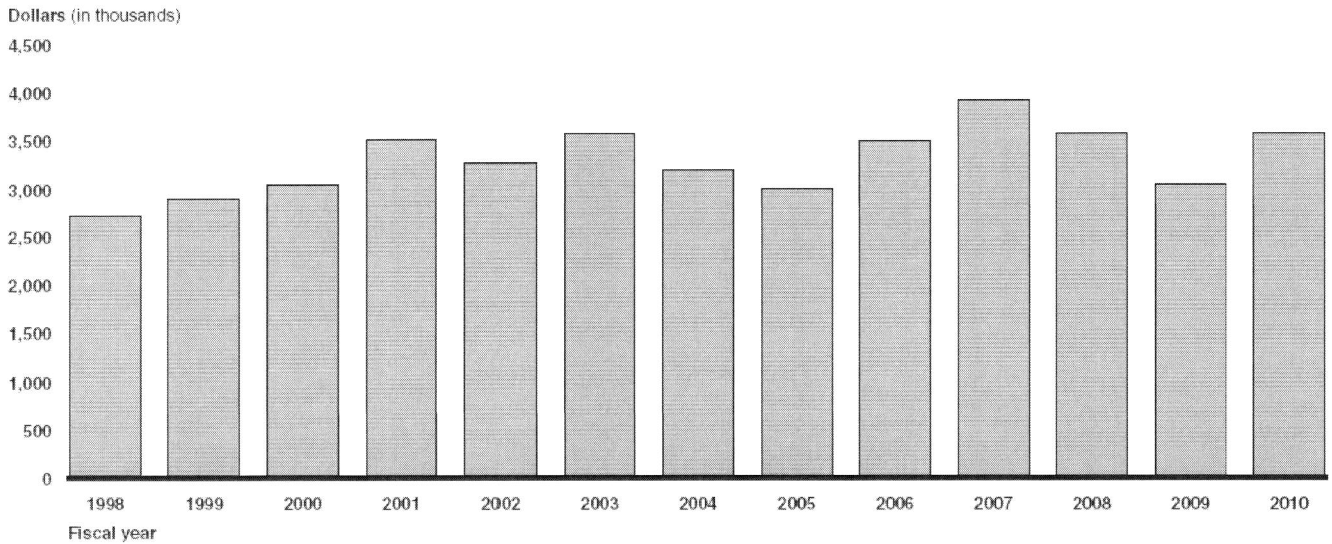

Figure 3. Estimated Environment and Natural Resources Division Attorney Costs for EPA Defensive Cases, Fiscal Year 1998 through Fiscal Year 2010.

On Average, Justice Spent at Least $3.3 Million a Year Defending EPA against Environmental Litigation in Fiscal Year 1998 through Fiscal Year 2010

Our analysis of data from Justice's Environment and Natural Resources Division found that from fiscal year 1998 through fiscal year 2010, Justice spent at least $3.3 million on average annually to defend EPA against environmental litigation, for a total of $43 million.[27] (The Environment and Natural Resources Division fiscal year 2010 budget was $110 million.) The U.S. Attorneys' Offices' database, however, does not contain information on attorney hours worked by case, which meant that we could not include the time these attorneys spent on each case in our estimate. According to Justice officials, however, the $3.3 million average per year represents the majority of Justice's time spent defending EPA each year, given that the U.S. Attorneys' Offices handle a small number of environment-related cases each year.[28] Overall, as shown in figure 3, annual costs increased by an average of about 3 percent each year from fiscal year 1998 through 2010, ranging from a low of $2.7 million in fiscal year 1998 to a high of $3.9 million in fiscal year 2007.

The Department of Justice Does Not Have a Standard Approach for Maintaining Key Data on Environmental Litigation Cases

Justice maintains separate, decentralized databases containing environmental case information and does not have a standard approach for collecting and entering data on these cases. Without a standard approach, it is it difficult to identify and summarize the full set of environmental litigation cases and costs managed by the department agencywide. Specifically, the department's Environment and Natural Resources Division and the U.S. Attorneys' Offices maintain different case management systems, and these systems do not use the same unique number to identify cases, making it possible to track cases within each component but not to align and merge cases from the two components. Because the U.S. Attorneys' Offices may assist the Environment and Natural Resources Division in certain case activities, a single case may appear in both systems, each with a different unique identifier. The only piece of data in both databases that can in practice be used to identify cases managed by both components is the court number, yet neither system has adopted the standard court number format used in the federal judiciary's Public Access to Court Electronic Records system, an electronic service that allows public access to case and docket information from federal appellate, district, and bankruptcy courts. According to an official of the Executive Office for U.S. Attorneys, the individual U.S. Attorneys' Offices may enter the court numbers in the specific formats used by the courts in their individual jurisdictions, although the official also said that there is no formal or written guidance for proper format of court numbers. Without such standard identifying numbers, it is difficult to identify a full and unduplicated list of environmental litigation cases and to derive descriptive statistics on costs, statute, or opposing parties. Because the department's data on environmental litigation cannot be reliably merged or aggregated to provide summary information on environmental cases, we had to use an iterative electronic and manual process to compile data from the two systems to conduct our review and identify the full set of environmental litigation cases and associated costs.

Moreover, not only are the two Justice databases separate, but the two agency components do not collect the same types of data on environmental cases. Specifically, the U.S. Attorneys' database does not collect data on the number of hours attorneys spend on an individual case or information on the statute under which a case is filed. As a result, it is impossible to gather complete data on all environmental litigation cases and costs from these databases. For example, we were unable to calculate the total number of hours that Justice attorneys worked on environmental cases—and hence, total costs of attorney time—because the U.S. Attorneys' time is not tracked by case.

By employing an iterative electronic and manual process to standardize the court numbers associated with all cases and matching cases from the two systems by these numbers, we were ultimately able to merge the two sets of data on environmental litigation cases managed by Justice's Environment and Natural Resources Division and the U.S. Attorneys' Offices for purposes of this report. Justice officials said, however, that they do not plan to change their approach to managing the data because they use the data in each system to manage individual cases, not to identify and summarize agencywide data on cases or trends. Officials said that their systems were designed for internal management purposes and not agencywide statistical tracking. Furthermore, while funds are spent to maintain the systems, officials indicated that the systems are old, and adding data fields or otherwise making changes to the systems may be technically infeasible or too costly. Justice officials said that the department previously sought to develop and implement a single case management system to gather common data agencywide, but the project was terminated in 2010 after a 2009 Office of the Inspector General report found that the project was more than 2 years behind the initial estimated completion date and that the project's total cost would be more than $18 million over budget. Because the two Justice components are not regularly required to merge and report their data in a systematic way, we are not making a recommendation regarding these data or systems.

On Average, Treasury Paid Successful Plaintiffs $1.8 Million Annually from the Judgment Fund over the Last 8 Years, and EPA Paid about $280,000 a Year over the Last 5 Years

In addition to Justice's costs of defending EPA, costs of litigation include payment of attorney fees and court costs to plaintiffs who prevail in lawsuits against EPA. As part of the payment process, Justice negotiated payment amounts with prevailing parties before finalizing the amount to be paid. For most of the claims under the 10 environmental statutes in this report, payments to successful plaintiffs were made from Treasury's Judgment Fund. Justice defended approximately 2,500 EPA-related cases filed from fiscal year 1995 through fiscal year 2010, but the number of environmental litigation cases from which plaintiffs received payments was small, representing about 8 percent of all cases.[29] In addition, EPA made a small number of payments for attorney fees and costs under the appropriate provision of EAJA.

Treasury's Judgment Fund Payments

From fiscal year 2003—the first year for which Treasury's Judgment Fund data are available—through fiscal year 2010, Treasury made, on average, 26 payments totaling $1.8 million per year for EPA-related environmental cases. The average Judgment Fund payment was $68,600 per payment. Treasury paid a total of about $14.2 million out of its Judgment Fund to prevailing plaintiffs for attorney fees and costs related to these cases (see fig. 4). The largest share of monies (46 percent) were made in cases filed by national environmental groups, followed by monies paid for cases filed by local environmental and citizens' groups (29 percent). The payments ranged from as little as $145, to the administrator of a law school clinic for a Clean Air Act suit, to as much as $720,000, to a private law firm for a Clean Water Act suit. According to Justice officials, payments are made either to the plaintiff or to the plaintiff's attorneys. Appendix III lists payments from Treasury's Judgment Fund for the environmental statutes in our review.

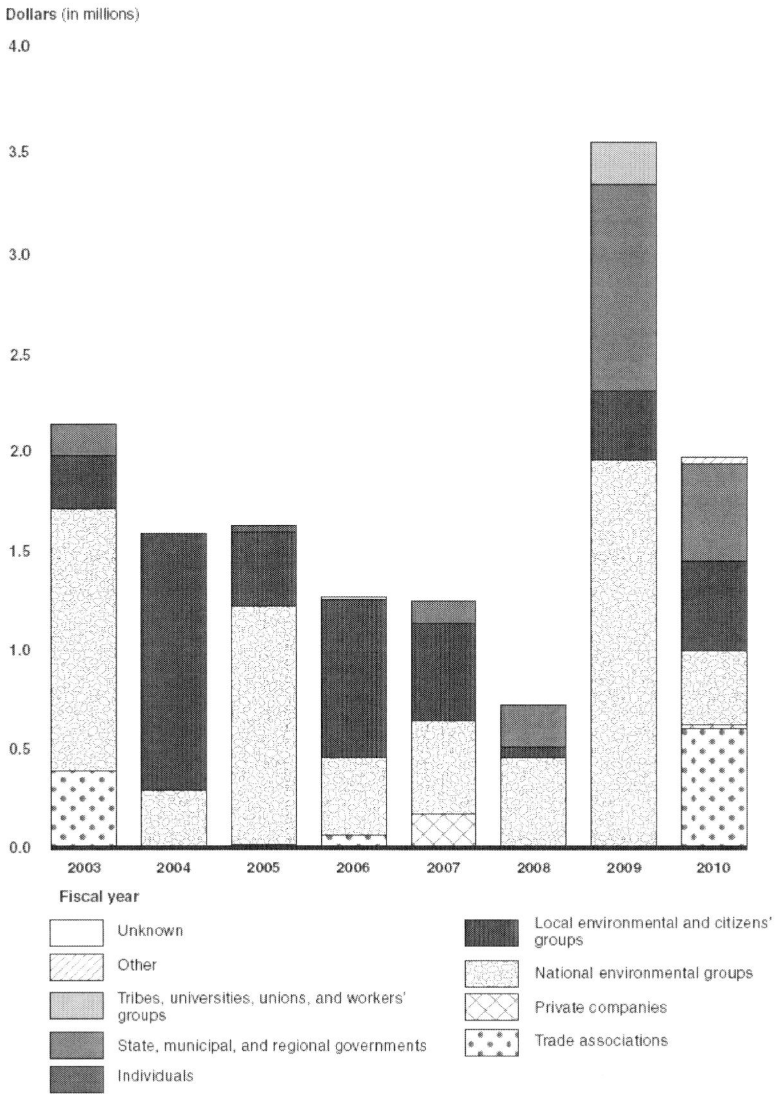

Source: GAO analysis of Treasury data.

Note: We omitted Superfund cases because we could not determine whether Treasury's **Judgment Fund** payments were for attorney fees and court costs or for reimbursements of site cleanups, which are not within the scope of our work.

Figure 4. Treasury's **Judgment Fund** Payments, Fiscal Year 2003 through Fiscal Year 2010.

Fluctuations in annual payments may occur, according to Justice officials, because payments to plaintiffs can be made several years after a case is completed, in part because Justice attempts to negotiate settlements of attorney fee claims before seeking a determination by the courts regarding claims that cannot be settled. Officials said that through this -process of negotiation, the department pays plaintiffs, in the majority of cases, an amount that is much lower than requested.

To determine attorney fees for each case, Justice considers, among other things, documentation by the plaintiff, including such factors as (1) the number of hours the plaintiff's attorneys spent on the case, which must be documented by the plaintiff; (2) the job description of the person spending time on the case (e.g., the costs for a paralegal and a lead counsel would be very different); (3) the specific tasks performed; and (4) applicable law in the jurisdiction, such as limits on hourly attorney fees or total amounts that courts have approved in the past.[30]

Although Justice may conclude that the hours are justified, fees may still be denied because of court precedent. Each time fees are negotiated, depending on the amount, Justice's Assistant Attorney General or the relevant Environment and Natural Resources Division Section Chief must approve the result, pursuant to applicable regulations and delegations.

EPA Payments

From fiscal year 2006—the first year for which EPA specifically tracked the payments by type of claim—through fiscal year 2010, EPA made 14 payments, totaling $1.4 million, for attorney fees and other costs under EAJA. EPA made an average of 2.8 payments per fiscal year, with an average payment of about $100,000. On average, EPA paid about $280,000 per year. The largest share of the monies (61 percent) went to payments for claims filed by local environmental groups, followed by monies (23 percent) for claims filed by national environmental groups. Although workers' groups filed comparatively few lawsuits, one such group did receive a single payment of $230,000 in fiscal year 2010 (see fig. 5).

The EPA payments ranged from $1,179, which was paid to an individual for a Clean Water Act suit in 2010, to $472,967, which was paid to an environmental group for two Clean Water Act suits, including one appeal. Appendix III contains a list of payments by payee.

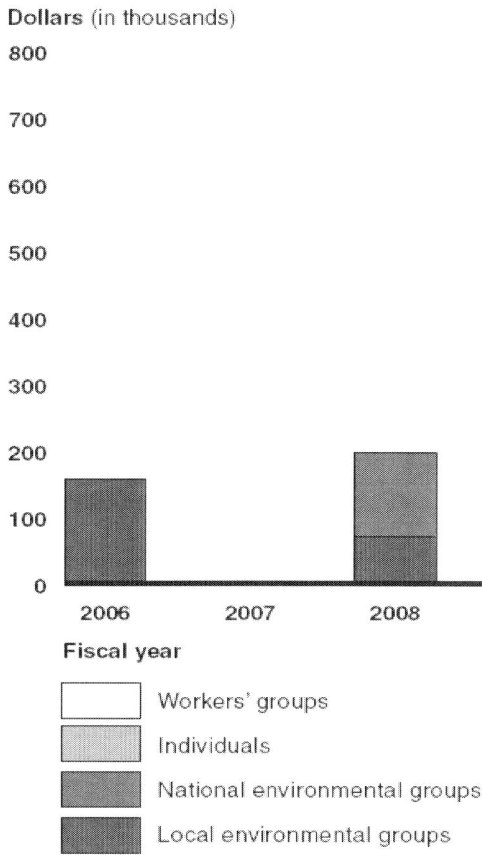

Source: GAO analysis of Treasury data.
Note: EPA made payments in calendar year 2007 but not in fiscal year 2007.

Figure 5. Amount Paid by EPA under EAJA for Environmental Cases, Fiscal Year 2006 through Fiscal Year 2010.

AGENCY COMMENTS AND OUR EVALUATION

We provided a draft of this report to EPA, Justice, and Treasury for their review and comment. EPA did not provide comments, and Justice and Treasury had technical comments, which we incorporated as appropriate.

David C. Trimble
Director, Natural Resources and Environment

APPENDIX I: OBJECTIVES, SCOPE, AND METHODOLOGY

This report describes (1) trends, if any, in environmental lawsuits against the Environmental Protection Agency (EPA) from fiscal year 1995 through fiscal year 2010, as well as stakeholders' views of factors affecting any trends, and (2) Justice's recent costs for representing EPA in defensive environmental lawsuits and the federal government's recent payments to plaintiffs.

To examine the changes over time to EPA's environmental litigation caseload, we obtained and analyzed data on lawsuits filed against the agency from databases maintained by two components within the Department of Justice—the Case Management System database maintained by Justice's Environment and Natural Resources Division and the Legal Information Office Network System database maintained by Justice's U.S. Attorneys' Offices. We obtained and analyzed data from these databases for lawsuits:

- filed in federal court from fiscal year 1995 through fiscal year 2010 (Oct. 1, 1994, through Sept. 30, 2010);
- in which EPA was the lead defendant, excluding cases in which EPA was a defendant but the lead defendant identified by Justice was another agency, such as the U.S. Army Corps of Engineers;
- brought under 10 major environmental statutes implemented by or applying to EPA, including the Clean Air Act; Clean Water Act; Safe Drinking Water Act; Resource Conservation and Recovery Act; Comprehensive Environmental Response, Compensation, and Liability Act (Superfund); Emergency Planning and Community Rightto-Know Act; Federal Insecticide, Fungicide, and Rodenticide Act; Federal Food, Drug, and Cosmetic Act; Toxic Substances Control Act; and the Endangered Species Act as it applies to EPA.

We excluded cases filed under the National Environmental Policy Act (NEPA) because these cases are managed by a number of sections within the Environment and Natural Resources Division, and because, according to Justice officials, few cases are filed under NEPA with EPA as the lead defendant. We also excluded the Freedom of Information Act, Discrimination in Federal Employment Act, Fair Labor Standards Act, and other generally applicable laws because the intent was to focus on challenges to EPA's core work in implementing environmental laws. Likewise, we excluded bankruptcy cases and cases heard in state court unless they were moved to federal court.

To determine if the data were reliable for our purposes, we checked them for completeness and legitimate values. When we were uncertain of the data's accuracy, we requested clarification from the source of the data. Within each database, we checked for duplicate records and either combined data across records into one record or removed unnecessary records. To compile a list of all cases of EPA lawsuits, we needed to identify duplicate cases across the two databases. Because the common field in the two systems—court number—is not kept in the same format, it was necessary for us to standardize court numbers into one format. To do so, we used the standard court number format used in the federal judiciary's Public Access to Court Electronic Records (PACER) system, an electronic public-access service that allows users to obtain case and docket information from federal appellate, district, and bankruptcy courts. After electronically processing reports of matched and unmatched cases, we conducted extensive manual review of the data to (1) confirm that matched cases from the two databases were in fact the same and (2) identify cases that were the same but were still not found with the electronic process. Manual checks of selected individual court cases were performed using the PACER system to correct information, such as EPA's role in the case, the names of plaintiffs, and court numbers.

We analyzed selected data elements—such as plaintiffs' names, filing and disposition dates, and relevant statute—over time to identify any trends in litigation. We also used the data on plaintiffs to identify categories of plaintiffs that have filed suit against EPA. To do this analysis, we used a process known as content analysis, searching national databases for information on each plaintiff and then using this information to code the plaintiffs according to rules developed by our internal team of analysts and specialists in program evaluation methods. Our team created 13 categories into which plaintiffs were coded (see table 3).

We evaluated the reliability of our plaintiff categories using two pretests on simple random samples of 40 and 41 plaintiffs, respectively. A minimum of five analysts independently coded the samples to ensure they had a common understanding of the categories and made the same coding decisions. For each pretest, we estimated the analysts' agreement rates adjusted for the possibility of agreement by chance. These "kappa" statistics estimate the reliability of each category. In the first pretest, the analysts agreed 74 percent of the time across all categories and 71 to 91 percent of the time for the individual categories other than "unknown," using the combined category of "local environmental and citizens' groups." On the basis of the results of the first pretest, we refined the definitions of the categories and conducted the second pretest. In the subsequent pretest, the analysts agreed 87 percent of the time across all categories and 84 to 95 percent of the time for the individual categories other than "unknown" and "other." These agreement rates suggested that the analysts could reliably classify the plaintiffs according to common standards in academic literature on intercoder agreement. Classifying the plaintiffs helped us quantify the number of cases brought each year against EPA by different types of groups.

Table 3. Categories of Plaintiffs

Plaintiff type	Description
Trade association	Represents numerous individual member for-profit companies and businesses
Private company	Privately owned business
Workers' group or labor union	Represents workers' interests; differs from associations
State or territory	State attorneys general, state environmental agencies, and other state government departments or agencies, including U.S. territories or their government offices
Municipal and regional government entity	Cities, counties, and towns, as well as local or regional public water authorities, among others
Tribe	Any Native American tribe or tribal organization
National environmental group	Environmental groups or associations working at the national level—often with one or more local offices or branches. For our purposes, "environment" meant a focus on conservation and improvement of any aspect of the outdoor environment, such as air orwater quality and protection of natural resources
Local environmental group	A group without a national parent group that works at the local and regional level
Citizens' group	Civic and social associations, as well as community groups, with an interest in citizen protection, health, or justice issues that are not identified as specifically focused on "environmental" issues
Individual	A private citizen not affiliated with any formal organization
University	Any group named as a college or university
Other	Any other plaintiffs that do not fit into the categories listed above
Unknown	A limited number of plaintiffs for which we were unable to find information, perhaps because those groups no longer existed or had changed their names

Source: GAO.

After validating the categories, we searched in public databases of organizations for information that would allow us to classify each plaintiff. We used the Nexis Encyclopedia of Associations and the Nexis Company Profile data systems, both of which identify organizations by North American Industry Classification System and Standard Industrial Classification. To the extent possible, we used these codes to classify plaintiffs. If these sources were not sufficient, we searched the Web pages of each organization for self-reported information. For the "individual" category of plaintiff, we confirmed through court records that those people were in fact suing as private individuals and not, for example, as mayors or attorneys general of a state.

Table 4. Stakeholders We Interviewed

Environmental groups
Center for Biological Diversity
Chesapeake Bay Foundation
Earthjustice
Environmental Integrity Project
Natural Resources Defense Council
Sierra Club
Trade associations
American Chemistry Council
American Forest and Paper Association
American Petroleum Institute
National Association of Homebuilders
National Association of Manufacturers
Utility Air Regulatory Group
State officials
California Attorney General's Office
Massachusetts Attorney General's Office
New York Attorney General's Office
Ohio Attorney General's Office
Texas Commission on Environmental Quality
Virginia Attorney General's Office
Other
National Association of Attorneys General
Widener University School of Law

Source: GAO.

In some cases, insufficient information was available in Justice's databases to determine a given plaintiff's identity. In such cases, we looked up the case in the PACER system. Six analysts conducted the content analysis of plaintiffs in the Case Management System and the Legal Information Office Network System. Discrepancies in coding were discussed, and agreement was reached among the analysts or resolved through a group analyst review.

To obtain stakeholder perspectives on environmental litigation trends and the factors that underlie them, we interviewed officials from EPA and Justice; representatives from the offices of five state attorneys general and one state environment department; representatives from six environmental groups; and six industry trade associations. We also spoke with a representative of the National Association of Attorneys General. Additionally, we interviewed one academic expert who has published extensively on environmental litigation in legal journals (see table 4). We selected these representatives on the basis of input from government officials and other interviewees. We asked the interviewees for their perspectives about factors that can affect trends in the types of lawsuits against EPA. We then performed a content analysis to group and summarize their responses. Not all stakeholders provided views on all issues, and statements from our sample of stakeholders cannot be generalized to all groups.

To determine Justice's costs for representing EPA in defensive environmental lawsuits and the government's payments to plaintiffs, we obtained data on three components of costs: (1) Justice's costs for its attorneys' time defending EPA, (2) payments for attorney and other costs from the Department of the Treasury's Judgment Fund for some cases that the government lost, and (3) payments for attorney and other costs by EPA for some cases that the government lost. For the first component, we obtained data from Justice on the number of cases per year that involved any of the 10 statutes in our scope, as well as the number of hours Justice attorneys spent working on these cases. To calculate costs, we multiplied the total hours worked in a given year by that year's average hourly pay rate—ranging from \$41 to \$66 per attorney for fiscal years 1998 through 2010—which we received from Justice. To adjust for uncompensated overtime, we reduced the reported annual hours the attorneys worked by 15 percent, an amount that Justice estimated represents overtime worked by its attorneys. To adjust the attorneys' salaries to include benefits and related agency overhead, we increased the attorneys' salaries by 84.3 percent, a factor that was provided to us by Justice on the basis of its actual 2009 costs. To ensure that attorney costs are comparable across years, we adjusted annual pay rates by applying the consumer price index for all urban consumers from the Department of Labor, Bureau of Labor Statistics, and inflated all pay rates to constant 2010 dollars. When we reported single payments, however, we did not adjust these figures to constant dollar figures.

To determine the second and third components of litigation costs—Treasury's Judgment Fund and EPA's payments to plaintiffs—we obtained and analyzed data from Treasury and EPA. First, we obtained and analyzed data from the Department of the Treasury's Judgment Fund Internet Claims System, which tracks the progress of plaintiffs' claims for Judgment Fund payments from the time they are sent to Treasury until the time they are paid. To identify data on payments related to the environmental statutes in our scope, we matched Treasury's data with data from Justice's two databases and eliminated payments that did not correspond with cases in our scope. When information was determined to be missing, we asked Treasury to provide us with additional information. In particular, we learned that Treasury's data included payments that were issued but were not cashed or were returned; we worked with Treasury to remove these payments to avoid counting these as actual payments and overrepresenting the amount paid from the Judgment Fund. We deleted Superfund cases because we were unable to discern from available information whether the Superfund-related payments were for attorney fees and court costs or for reimbursements of site cleanups, which is a different category of payment than what is within our scope. Similarly, to identify EPA payments to plaintiffs within our scope under the Equal Access to Justice Act, we obtained EPA data on payments made to plaintiffs and manually matched these cases with the cases in Justice's two databases. When certain case information was determined to be missing, we did additional research on these cases using PACER and corrected the data. Inconsistent formatting of key data elements produced significant problems for completing our analysis and required significant manual review by us and Justice. If we did not find the necessary information from available sources, we asked EPA to send us relevant portions of the internal voucher packages used to request payment.

We conducted this performance audit from June 2010 through July 2011, in accordance with generally accepted government auditing standards. Those standards require that we plan and perform the audit to obtain sufficient, appropriate evidence to provide a reasonable basis for our findings and conclusions based on our audit objectives. We believe that the evidence obtained provides a reasonable basis for our findings and conclusions based on our audit objectives.

APPENDIX II: KEY PROVISIONS AND HISTORICAL REPORTING UNDER THE EQUAL ACCESS TO JUSTICE ACT

In the United States, parties involved in litigation generally bear their own attorney fees and costs. For policy reasons, including encouraging citizens to bring suits to enforce the law, Congress has provided exceptions to this rule for cases brought under several statutes, such as the Civil Rights Act. In these instances, as well as some common-law exceptions, a prevailing plaintiff may seek awards of its attorney fees and court costs from the losing party.

Historically, the federal government had sovereign immunity from some of these exceptions, but in some instances, the statutes also waived sovereign immunity so that a court could award fees and costs against the federal government, as well as a private party. According to Justice, many of the key environmental statutes' provisions authorizing award of attorney fees and costs apply to the federal government. For example, EPA pays attorney fees under several provisions of the Clean Air Act and the Clean Water Act.[1]

Furthermore, in 1980, the Equal Access to Justice Act (EAJA) was enacted to waive sovereign immunity for the remaining statutes authorizing award of fees and costs, as well as to authorize the awarding of fees and costs in other cases. As the 1980 conference committee report for EAJA explains, the act's premise is that individuals, corporations, partnerships, and labor and other organizations did not seek review of or defend against unreasonable government actions because of the expense involved, which was compounded by the disparity in expertise and resources between the government and the individual or organization involved. EAJA was intended to help certain individuals, partnerships, corporations, and labor and other organizations by paying the attorney fees and other costs if the federal government brought an administrative or judicial action and lost because the action was not substantially justified. EAJA seeks to (1) encourage parties that are the subject of unreasonable federal government action to seek reimbursement for attorney fees and other costs, (2) restrain overzealous regulators, and (3) ensure that the government pays for the cost of refining and formulating public policy.[2]

EAJA authorizes the award of the following:

- Court costs of prevailing parties against the United States in any civil action. These costs may include fees for the clerk and marshal, reporter, printing, witnesses, copies, docket fees, and interpreters and court-appointed experts and may include an amount equal to the filing fees.

- Attorney fees and expenses against the United States of a prevailing party to the same extent as any other party, codified at 28 U.S.C. § 2412(b) and hereinafter referred to as "subsection b." That is, where there is a statutory or common-law exception that provides for award of fees to a prevailing party, such exceptions also apply to the federal government.[3] Regarding the 10 environmental statutes covered in this report, many of the relevant provisions under which EPA may be sued provide for award of such fees.[4] However, EAJA makes EPA subject to fee awards under all the environmental statutes' provisions authorizing courts to award attorney fees and expenses. According to Justice, many of the environmental suits against EPA involve provisions that authorize fee awards independent of EAJA, but a small number may fall into EAJA subsection b. A feature of this subsection is that it does not itself limit the eligibility of prevailing plaintiffs, nor expressly limit the hourly rate of attorney fees; however, the statute requires that the fees be "reasonable." Additionally, any award of fees made under this section is subject to any limitations that would apply to analogous awards against private parties, as may be provided by the underlying statute.
- Attorney fees and expenses of a prevailing party in cases even when no statutory or common-law exception exists to make a private defendant liable for such fees, unless the court finds that the position of the United States was substantially justified or that special circumstances make an award unjust. This subsection of EAJA, codified at 28 U.S.C. § 2412(d) and hereinafter referred to as "subsection d," authorized the award of these fees against the federal government in civil court actions,[5] while another subsection authorized the award of these fees in certain agency adjudications such as when a party files an appeal of an agency decision to the EPA Environmental Appeals Board. Two of the 10 laws covered in this report—the Federal Insecticide, Fungicide, and Rodenticide Act and the Federal Food, Drug, and Cosmetic Act—as well as some individual provisions of other statutes, do not authorize payment of fees to prevailing parties. Cases brought against EPA under these statutes and provisions, then, fall into EAJA subsection d. This subsection limits the prevailing plaintiff's eligibility to receive payment by defining an eligible party as, at the time the lawsuit is

filed, either an individual with a net worth below $2 million or a business owner or any partnership, corporation, association, local government, or organization with a net worth below $7 million and fewer than 500 employees. Tax-exempt nonprofit organizations and certain agricultural marketing cooperatives are considered parties regardless of net worth.

Payment of attorney fees by federal agencies under statutes independently authorizing awards against federal agencies and under subsection b are made from the Judgment Fund, which is a permanent, indefinite appropriation available to pay many money judgments against the United States.[6] Payment of attorney fees by federal agencies under subsection d is generally made from agency appropriations. Table 5 summarizes key attributes of the three authorizing situations under which EPA may pay fees and costs.

Table 5. Key Attributes of Attorney Fee Payment Provisions Relevant to Judicial Proceedings

Authorizing statute	Applicability	Limitations	Reporting requirement	Funding for payments
Various provisions of some environmental statutes	—	Varies	No	JudgmentFund
EAJA subsection b	Where statute or common law provides for award of fees against private party	No (any limitations in the statute(s) allowing fees to be awarded apply)	No	JudgmentFund
EAJA subsection d[a]	Where there is no requirement for payment of fees to private parties in law	Yes (hourly rates of attorney fees are limited and restrictions on plaintiffs eligibility for fees)	Yes (repealed)	Agency appropriations

Source: GAO.

[a] A separate section of EAJA, codified at 5 U.S.C. § 504, provides for similar payments in agency adjudications.

Originally, EAJA provided for governmentwide reporting on its use and cost. For judicial proceedings, EAJA required the Director of the Administrative Office of the U.S. Courts to report annually to Congress on EAJA court activity, including the number, nature, and amounts of awards; claims involved; and any other relevant information deemed necessary to aid Congress in evaluating the scope and effect of awards under the act. The responsibility for this reporting was transferred to the Attorney General in 1992. In addition, EAJA required the Chairman of the Administrative Conference of the United States to submit an EAJA report annually to Congress on administratively awarded fees and expenses. Then, in December 1995, the Federal Reports Elimination and Sunset Act of 1995 repealed the Attorney General's reporting requirement for fees and expenses awarded under EAJA and also discontinued reporting of governmentwide administrative awards of fees and costs under EAJA after fiscal year 1994. Currently, there are no statutory requirements in effect for agency or governmentwide reporting of payments made under EAJA for either administrative or judicial proceedings. According to officials from the Administrative Conference of the United States, the conference has begun to obtain and compile such information for fiscal year 2010, noting that there has been continued interest in Congress (including pending legislation) regarding data about payments under EAJA. Officials told us the conference has requested EAJA data from 50 government agency conference members, as well as a few additional agencies that had previously reported EAJA activity to the conference. The chairman plans to publish a report for fiscal year 2010 later in 2011.

We have previously reported certain governmentwide EAJA data, as well as data focused on selected agencies. In 1996, we reported data on the number of cases and amounts of awarded plaintiff attorneys' fees exceeding $10,000 against nine federal agencies for cases closed during fiscal years 1993 and 1994.[7] In 1998, we provided information on the history of EAJA, the extent to which one provision of the act was used governmentwide from 1982 to1994, and the provision's use by the Department of Labor and associated agencies.[8] The governmentwide data showed, among other things, that the Departments of Health and Human Services and of Veterans Affairs accounted for most EAJA payments in court proceedings, under the provision that applies when the substantive law does not authorize award of attorney fees and costs.[9]

Table 6. Treasury's Judgment Fund Payments, Fiscal Year 2003 through Fiscal Year 2010

Date sent	Case title	Payee(s)	Payment amount[a]	Statute
9/29/2010	Louisiana Environmental Action Network v. Jackson	Administrators of the Tulane Educational Fund	$5,000.00	Clean Air Act
9/27/2010	Sierra Club v. EPA	McGillivray Westerberg	3,223.38	Clean Air Act
9/22/2010	Heal the Bay Inc. v. EPA	Natural Resources Defense Council	17,051.00	Clean Water Act
9/21/2010	Sierra Club v. EPA	Earthjustice Attorney	350.00	Clean Air Act
8/19/2010	National Pork Producers Council and American Farm Bureau v. EPA	Waterkeeper Alliance	25,447.53	Clean Water Act
8/19/2010		Natural Resources Defense Council	69,552.47	Clean Water Act
8/17/2010	Sierra Club v. Jackson	IOLTA Law Office of Robert Ukeiley, PSC[b]	8,000.00	Clean Air Act
8/3/2010	Sierra Club v. EPA	Earthjustice Attorney	11,019.57	Clean Air Act
7/14/2010	Sierra Club v. EPA	McGillivray Westerberg	2,624.71	Clean Air Act
7/13/2010	Northwest Environmental Defense Center et al. v. U.S. Environmental Protection Agency et al.	Pacific Environmental Advocacy Center Attorney Escrow Account/IOLTA	40,000.00	Endangered SpeciesAct
7/8/2010	Environmental Integrity Project v. Jackson	Vermont Law School Inc.	8,000.00	Clean Air Act
6/25/2010	Natural Resources Defense Council v. EPA	Earthjustice Attorney	198,700.00	Clean Air Act
6/22/2010	Florida Wildlife Federation v. EPA	Earthjustice Florida	198,997.00	Clean Water Act
6/11/2010	WildEarth Guardians v. EPA	IOLTA Law Office of Robert Ukeiley, PSC	16,000.00	Clean Air Act

Date sent	Case title	Payee(s)	Payment amount[a]	Statute
5/25/2010	The National Cotton Council of America v. EPA	Western Environmental Law Center	500,000.00	Clean Water Act
5/24/2010	State of New Jersey v. EPA	Clean Air Act Task Force	400,000.00	Clean Air Act
3/12/2009		Earthjustice Attorney	2,710.96	Clean Air Act
5/19/2010	State of New York v. EPA	Earthjustice Attorney	95,000.00	Clean Air Act
11/16/2009		Earthjustice Attorney	1,476.21	Clean Air Act
4/28/2010	WildEarth Guardians v. EPA	WildEarth Guardians	3,520.00	Clean Air Act
4/28/2010	Comite Civico Del Valle Inc. v. Jackson	Law Office of Gideon Kracov	33,000.00	Clean Air Act
4/28/2010	Association of Irritated Residents v. EPA	Center on Race, Poverty, and the Environment	16,500.00	Clean Air Act
3/23/2010	WildEarth Guardians v. EPA	WildEarth Guardians	4,588.50	Clean Air Act
3/22/2010	WildEarth Guardians v. Jackson	Robert Ukeiley, PSC IOLTA	22,420.00	Clean Air Act
3/10/2010	Mossville Environmental Action Now v. EPA	Earthjustice Attorney	8,000.00	Clean Air Act
3/4/2010	WildEarth Guardians v. Jackson	Earthjustice	16,000.00	Clean Air Act
3/3/2010	Desert Rock and Dine Power Authority v. EPA	Ater Wynne, LLP IOLTA and Oregon Law Foundation	20,971.12	Clean Air Act
3/3/2010	Sierra Club v. EPA	Tom Neltner	65,489.00	Toxic SubstancesControl Act
3/3/2010	New York Coalition to End Lead Poisoning v. EPA	Northern Manhattan Improvement Corporation	62,393.00	Toxic SubstancesControl Act
2/25/2010	Environmental Integrity Project v. Jackson	Environmental Integrity Project	6,312.83	Clean Air Act
2/24/2010	Sierra Club v. Jackson	McGillivray Westerberg	2,847.64	Clean Air Act
1/27/2010	Sierra Club v. Jackson	Robert Ukeiley, PSC IOLTA	4,000.00	Clean Air Act

Table 6. (Continued)

Date sent	Case title	Payee(s)	Payment amount[a]	Statute
1/14/2010	Florida Clean Water Act Network v. EPA	Law Office of David A. Ludder PLLC	13,634.50	Clean Water Act
12/17/2009	Sierra Club v. Johnson	McGillivray Westerberg	4,901.26	Clean Air Act
12/8/2009	Sierra Club v. EPA	Earthjustice Attorney	3,314.48	Clean Air Act
11/24/2009	Colorado Citizens Against Toxic Waste v. EPA	Colorado Lawyer Trust	27,427.50	Clean Air Act
11/16/2009	Association of Irritated Residents v. EPA	Center on Race, Poverty, and the Environment	11,500.00	Clean Air Act
11/13/2009	Sierra Club v. EPA	Earthjustice Attorney	1,150.00	Clean Air Act
10/30/2009	Sierra Club v. Stephen Johnson	Sierra Club	21,311.00	Clean Air Act
10/8/2009	Sierra Club v. EPA	Reed Zars	31,140.00	Clean Air Act
10/5/2009	Sierra Club v. EPA	McGillivray Westerberg	2,057.54	Clean Air Act
9/28/2009	American Farm Bureau Federation v. EPA	Earthjustice Attorney	209,867.00	Clean Air Act
9/22/2009		Earthjustice Attorney	1,360.44	Clean Air Act
9/24/2009	Center for Biological Diversity v. Johnson	Robert Ukeiley, PSC	22,500.00	Clean Air Act
9/22/2009	WildEarth Guardians v. Johnson	Robert Ukeiley, PSC	4,884.00	Clean Air Act
8/6/2009	Environmental Integrity Project v. EPA	Earthjustice Main Account	67,587.80	Clean Air Act
8/5/2009	Sierra Club v. Johnson	Sierra Club	5,000.00	Toxic SubstancesControl Act
6/23/2009	Sierra Club v. EPA	McGillivray Westerberg	2,787.06	Clean Air Act
6/16/2009	Friends of Pinto Creek v. EPA	Western Mining Action	162,347.91	Clean Water Act
6/12/2009	State of North Carolina v. EPA	Minnesota Power	100,000.00	Clean Air Act

Date sent	Case title	Payee(s)	Payment amount[a]	Statute
6/11/2009	Environmental Integrity Project v. EPA	Environmental Integrity Project	4,500.00	Clean Air Act
5/11/2009		Environmental Integrity Project	4,500.00	Clean Air Act
5/4/2009	Natural Resources Defense Council v. EPA	Earthjustice Attorney	163,500.00	Clean Air Act
5/1/2009	Rocky Mountain Clean Air Action v. Johnson	John M. Barth	8,526.02	Clean Air Act
4/2/2009	Natural Resources Defense Council v. EPA	New York State Department of Law	144.90	Clean Water Act
3/25/2009		Patterson Belknap Webb & Tyler, LLP	800.40	Clean Water Act
3/24/2009		Patterson Belknap Webb & Tyler, LLP	720,000.00	Clean Water Act
3/17/2009	Natural Resources Defense Council v. EPA	Natural Resources Defense Council	117,125.00	Clean Water Act
1/5/2009		Natural Resources Defense Council	628.70	Clean Water Act
2/17/2009	Northwest Environmental Advocates v. EPA	Northwest Environmental Advocates	60,000.00	Clean Water Act
2/10/2009	Commonwealth of Massachusetts v. EPA	Commonwealth of Massachusetts	127,555.00	Clean Air Act
1/23/2009		Sierra Club	93,205.00	Clean Air Act
1/12/2009		Lisa Heinzerling	100,000.00	Clean Air Act
12/15/2008		State of California Air Resources Board	111,277.00	Clean Air Act
12/15/2008		International Center for Technology Assessment	81,438.00	Clean Air Act
12/15/2008		City of New York	28,841.00	Clean Air Act
12/12/2008		Michael C. MacCracken	11,045.00	Clean Air Act

Table 6. (Continued)

Date sent	Case title	Payee(s)	Payment amount[a]	Statute
12/12/2008		Natural Resources Defense Council	54,002.00	Clean Air Act
12/12/2008		Earthjustice Main Account	92,637.00	Clean Air Act
2/5/2009	Communities for a Better Environment v. EPA	IOLTA Law Office of Robert Ukeiley, PSC	65,000.00	Clean Air Act
2/2/2009	Natural Resources Defense Council v. Johnson	Los Angeles County Flood Control District	201,000.00	Clean Water Act
1/14/2009		National Association of Clean Water Agencies	95,000.00	Clean Water Act
1/14/2009		Natural Resources Defense Council	498,000.00	Clean Water Act
1/5/2009	Association of Irritated Residents v. EPA	Center on Race, Poverty, and the Environment	33,815.00	Clean Air Act
12/31/2008		Natural Resources Defense Council	10,751.00	Clean Air Act
12/29/2008	South Coast Air Quality Management District v. EPA	Earthjustice Main Account	240,000.00	Clean Air Act
8/1/2008	Sierra Club v. EPA	Garvey McNeil & McGillivray	7,250.00	Clean Air Act
5/22/2008	Rocky Mountain Clean Air Action v. EPA	Robert Ukeiley, PSC	25,000.00	Clean Air Act
5/13/2008	Sierra Club V. EPA	Earthjustice Attorney	110,000.00	Clean Air Act
5/1/2008	Rocky Mountain Clean Air Action v. Johnson	Robert Ukeiley, PSC	5,283.28	Clean Air Act
3/20/2008	Sierra Club v. EPA	A.G. Edwards and Sons	7,843.00	Clean Air Act
2/26/2008	State of New York v. EPA	Earthjustice Inc.	198,000.00	Clean Air Act

Date sent	Case title	Payee(s)	Payment amount[a]	Statute
9/14/2006		Earthjustice Legal Services	2,745.44	Clean Air Act
2/22/2008	Center for Biological Diversity v. Johnson	Robert Ukeiley, PSC	55,145.00	Clean Air Act
2/22/2008	Sierra Club v. EPA	Garvey McNeil & McGillivray	6,500.00	Clean Air Act
2/22/2008	Environmental Defense Inc. v. EPA	Sierra Club	945.69	Clean Air Act
2/11/2008	Public Employees for Environmental Responsibility v. Johnson	Public Employees for Environmental Responsibility	40,000.00	Toxic SubstancesControl Act
12/5/2007	Sierra Club v. EPA	Sierra Club	3,454.66	Clean Air Act
12/4/2007	Sierra Club v. EPA	Account for Reed Zars	14,360.00	Clean Air Act
12/4/2007	Our Children's Earth Foundation v. EPA	Golden Gate University School of Law	18,000.00	Clean Air Act
11/29/2007	South Coast Air Quality Management District v. EPA	Earthjustice Attorney	2,353.62	Clean Air Act
11/27/2007	Sierra Club v. Johnson	McNeil & McGillivray, S.C. IOLTA	4,898.00	Clean Air Act
11/27/2007	State of New Jersey v. EPA	New Jersey Department of Environmental Protection	3,700.00	Clean Air Act
11/27/2007	Natural Resources Defense Council v. EPA	Natural Resources Defense Council	2,077.17	Clean Air Act
10/26/2007	American Lung Association of Metropolitan Chicago v. EPA	Environmental Law and Policy Center	3,075.00	Clean Air Act
10/25/2007	People for the State of Illinois v. EPA	State of Illinois	7,621.00	Clean Air Act
10/25/2007	Citizens Against Ruining the Environment v. EPA	Chicago Legal Clinic	3,075.00	Clean Air Act
10/18/2007	Sierra Club v. EPA	Earthjustice Attorney Client Trust Fund	190,000.00	Clean Air Act

Table 6. (Continued)

Date sent	Case title	Payee(s)	Payment amount[a]	Statute
9/27/2007	Natural Resources Defense Council v. EPA	Earthjustice	4,472.74	Clean Air Act
9/1/2007	Sierra Club v. EPA	Earthjustice Attorney Client Trust Account	1,630.18	Clean Air Act
8/22/2007	Rocky Mountain Clean Air Act Act Action v. Johnson	Robert Ukeiley, PSC	9,536.00	Clean Air Act
8/21/2007	Safe Air for Everyone v. EPA	Earthjustice Main Account	133,512.05	Clean Air Act
8/20/2007		Earthjustice Main Account	974.30	Clean Air Act
8/13/2007	Sierra Club v. EPA	Sierra Club	25,000.00	Toxic SubstancesControl Act
8/8/2007	Baykeeper v. EPA	Environmental Advocates Attorney Client Trust Account/IOLTA	125,000.00	Clean Water Act
7/19/2007	Idaho Conservation League v. Johnson	Robert Ukeiley, PSC	4,945.00	Clean Air Act
6/23/2007	Coralations v. EPA	Mid-Atlantic Environmental Law Center	95,388.40	Clean Water Act
5/22/2007	Sierra Club and Coosa River Basin Initiative v. EPA	Georgia Center for Law and the Public Interest	12,000.00	Clean Air Act
5/4/2007	Honeywell International Inc. v. EPA	Gomar National Industries	150,000.00	Clean Air Act
11/26/2004		Sidley Austin Brown & Wood LLP	956.74	Clean Air Act
4/24/2007	State of New York v. EPA	Mid-Atlantic Environmental Law Center	5,000.00	Clean Air Act
3/10/2007		Earthjustice Legal Services Trust	2,374.80	Clean Air Act

Date sent	Case title	Payee(s)	Payment amount[a]	Statute
12/7/2006		Earthjustice Legal Services Trust	98,837.65	Clean Air Act
4/13/2007	Center for Biological Diversity v. Christie Whitman et al.	Michael W. Graf	405,000.00	Endangered SpeciesAct
2/28/2007	Elmwood Park et al. v. Stephen Johnson et al.	Rutgers Environmental Law Clinic	1,900.00	Clean Air Act
1/17/2007	American Canoe Association Inc. v. EPA	John M. Simpson	2,398.00	Clean Water Act
11/20/2006	Medical Advocates for Healthy Air v. EPA	Earthjustice Main	83,134.11	Clean Air Act
11/15/2006	Kentuckians for the Commonwealth v. EPA	Robert Ukeiley, PSC	16,092.00	Clean Air Act
9/6/2006	Sierra Club v. EPA	Earthjustice Legal Services	67,550.00	Clean Air Act
8/30/2006	South Jersey Environmental v. Stephen Johnson, EPA	Eastern Environmental Law Center	1,871.00	Clean Air Act
7/31/2006	Louisiana Environmental Action Network v. Johnson	Administrators of the Tulane Educational Fund	1,306.50	Clean Air Act
7/12/2006	Environmental Integrity Project v. EPA	Earthjustice Legal Services	93,000.00	Clean Air Act
3/30/2006		Earthjustice Attorney Client Trust Account	1,732.21	Clean Air Act
6/20/2006	Washington Toxics Coalition et. al. v. EPA	Earthjustice Attorrney Client Trust Account	625,602.40	Endangered SpeciesAct
5/26/2006	South Jersey Environmental Justice Alliance v. Johnson	Eastern Environmental Law Center	5,502.00	Clean Air Act
5/4/2006	Sierra Club v. Johnson	Robert Ukeiley, PSC	6,800.00	Clean Air Act
4/26/2006	Our Children's Earth Foundation v. EPA	Our Children's Earth Legal Fund	14,123.34	Clean Air Act
4/21/2006	Sierra Club v. EPA	FBO Georgia Center for Law in the Public Interest	45,000.00	Clean Air Act
4/19/2006	Natural Resources Defense Council v. EPA	Natural Resources Defense Council	140,000.00	Endangered SpeciesAct

Table 6. (Continued)

Date sent	Case title	Payee(s)	Payment amount[a]	Statute
3/23/2006	Sierra Club V. EPA	Sierra Club	3,036.66	Clean Air Act
10/27/2005	Mossville Environmental Action Now v. EPA	Earthjustice Attorney	94,500.00	Clean Air Act
10/25/2005	Sierra Club v. Johnson	Sierra Club	11,981.86	Clean Air Act
10/25/2005	Blue Skies Alliance v. Leavitt	Marc Chytillo	45,000.00	Clean Air Act
10/7/2005	Sailors Inc. and Mississippi River Revival v. EPA	Jerry L. Anderson	11,000.00	Clean Water Act
9/19/2005	Environmental Defense v. Johnson	Southern Environmental Law Center	31,000.00	Clean Air Act
9/19/2005	Sierra Club v. Leavitt	Earthjustice	32,923.38	Clean Air Act
9/8/2005	National Wildlife Federation v. EPA	National Wildlife Federation	62,128.00	Clean Air Act
9/1/2005	Center for Biological Diversity et al. v. Environmental Protection Agency	Save Our Springs Alliance	53,000.00	Endangered SpeciesAct
8/17/2005	Waterkeeper Alliance v. EPA	Waterkeeper Alliance Inc.	415,145.00	Clean Water Act
6/29/2005		Crowell & Moring Master	220,141.34	Clean Water Act
7/26/2005	Martha Vigil v. EPA	Arizona Center for Law in the Public Interest	30,957.75	Clean Air Act
7/14/2005	American Lung Association of Metropolitan Chicago v. EPA	Sierra Club	3,570.00	Clean Air Act
7/13/2005		Environmental Law and Policy Center	24,585.00	Clean Air Act
6/22/2005		Chicago Legal Clinic Inc. Clients' Fund	18,480.00	Clean Air Act
6/9/2005	American Lung Association V. EPA	Earthjustice	18,214.00	Clean Air Act

Date sent	Case title	Payee(s)	Payment amount[a]	Statute
2/26/2004		Earthjustice	23,000.00	Clean Air Act
6/9/2005	American Lung Association v. Whitman	Earthjustice	6,650.00	Clean Air Act
12/30/2003		Earthjustice	78,500.00	Clean Air Act
5/12/2005	Bluewater Network v. EPA	Earthjustice Attorney Client Trust Account	45,406.21	Clean Air Act
4/28/2005	Sierra Club v. Whitman	Earthjustice	11,800.00	Clean Air Act
4/7/2005	Glynn Environmental Coalition Inc. v. EPA	Legal Environmental Assistance Foundation	7,908.64	Clean Air Act
3/30/2005	Sierra Club v. EPA	Earthjustice	53,920.80	Clean Air Act
6/17/2004		Earthjustice	1,347.96	Clean Air Act
3/25/2005	New York Public Interest Research Group v. Whitman	Morningside Heights Legal Services	35,000.00	Clean Air Act
3/25/2005	Environmental Defense v. Horinko	Earthjustice	22,276.00	Clean Air Act
2/20/2005	Louisiana Environmental Action Network v. Leavitt	The Administrators of the Tulane Educational Fund	1,641.50	Clean Air Act
1/25/2005	Missouri Coalition for the Environment v. Leavitt	Missouri Coalition for the Environment	88,198.79	Clean Water Act
12/29/2004	Environmental Defense v. EPA	Reed Zars	49,430.00	Clean Air Act
12/11/2004	Louisiana Environmental Action Network v. EPA	The Administrators of the Tulane Educational Fund	129,066.99	Clean Air Act
10/27/2004	Sierra Club v. EPA	Earthjustice	91,000.00	Clean Air Act
11/18/2003		Earthjustice	1,006.40	Clean Air Act
9/30/2004	Our Children's Earth Foundation v. EPA	Golden Gate University School of Law	30,000.00	Clean Air Act
9/30/2004	Sierra Club v. EPA	FBO Georgia Center for Law in the Public Interest	31,132.60	Clean Air Act

Table 6. (Continued)

Date sent	Case title	Payee(s)	Payment amount[a]	Statute
9/25/2004	Our Children's Earth Foundation v. EPA	Environmental Advocates	70,000.00	Clean Water Act
7/26/2004	Environmental Defense Center v. EPA	Paul Weiss Rifkind Wharton & Garrison	219,000.00	Clean Water Act
7/6/2004	Riverkeeper Inc. v. EPA	Riverkeeper Inc.	98,000.00	Clean Water Act
6/16/2004	Environmental Defense Fund and Natural Resources Defense Council v. EPA	Robert E. Yuhnke & Associates	27,000.00	Clean Air Act
6/8/2004	Sierra Club v. Leavitt	FBO Georgia Center for Law in the Public Interest	20,000.00	Clean Air Act
5/10/2004	Our Children's Earth Foundation v. EPA	Golden Gate University School of Law	4,500.00	Clean Air Act
4/30/2004	Medical Advocates for Healthy Air v. Whitman	Earthjustice Legal Defense Fund	199,622.98	Clean Air Act
4/12/2004	Sierra Club and New York Public Interest Research Group Inc v. EPA	Earthjustice	49,500.00	Clean Air Act
4/6/2004	Save the Valley Inc. v. EPA	Save the Valley Inc.	45,600.64	Clean Water Act
3/24/2004	Association of Irritated Residents v. EPA	Earthjustice Attorney	50,620.00	Clean Air Act
2/17/2004		California Rural Legal Assistance Foundation	5,320.50	Clean Air Act
3/22/2004	Our Children's Earth Foundation and Sierra Club v. EPA	Reed Zars	155,000.00	Clean Air Act
2/27/2004	Defenders of Wildlife et al. v. United States Environmental Protection Agency	Belin and Sugarman	26,526.96	Endangered SpeciesAct
2/5/2004	Louisiana Environmental Action Network v. EPA	Administrators of the Tulane Educational Fund	3,200.00	Clean Air Act

Date sent	Case title	Payee(s)	Payment amount[a]	Statute
2/3/2004	Bluewater Network v. EPA	Earthjustice	85,542.00	Clean Air Act
12/15/2003	International Center for Technology Assessment v. Whitman	International Center for Technology Assessment	29,000.00	Clean Air Act
11/5/2003	Southern Organizing Committee for Economic and Social Justice v. EPA	Southern Environmental Law Center	120,000.00	Clean Air Act
8/22/2003	Sierra Club v. Whitman	Georgia Center for Law in the Public Interest	14,464.48	Clean Air Act
7/25/2003	Kansas Natural Resource Council Inc. v. Browner	John M. Simpson PC	38,270.00	Clean Water Act
7/25/2003		Charles Benjamin Inc.	17,925.00	Clean Water Act
6/30/2003	New York Public Interest Research Group v. Whitman	Earthjustice	15,251.52	Clean Air Act
6/23/2003	Sierra Club v Whitman	Earthjustice	22,490.00	Clean Air Act
6/19/2003	Juanita Stewart v. Whitman	Administrators of the Tulane Educational Fund	150.00	Clean Air Act
6/18/2003	New York Public Interest Research Group v. EPA	New York Public Interest Research Group	3,004.90	Clean Air Act
5/5/2003	Sierra Club v. EPA	Earthjustice	145,669.00	Clean Air Act
10/17/2002		Earthjustice Legal Defense Fund	726.78	Clean Air Act
4/30/2003	Sierra Club and Missouri Coalition for the Environment v. EPA	Green Hennings & Henry	66,146.38	Clean Air Act
4/30/2003	American Corn Growers Association v. EPA	Western Fuels Association Inc.	22,000.00	Clean Air Act
4/30/2003		Center for Energy and Economic Development	182,000.00	Clean Air Act
4/24/2003	Sierra Club v. EPA	Earthjustice	41,584.00	Clean Air Act

Table 6. (Continued)

Date sent	Case title	Payee(s)	Payment amount[a]	Statute
4/21/2003	Sierra Club v. EPA	Marc Chytilo	77,000.00	Clean Air Act
4/2/2003	South Coast Air Quality Management District v. EPA	Los Angeles County Treasurer	102,000.00	Clean Air Act
3/4/2003	Communities for a Better Environment v. EPA	Environmental Law and Justice Clinic	36,491.00	Clean Air Act
1/28/2003	Natural Resources Defense Council v. Browner	Earthjustice	14,909.00	Clean Air Act
12/21/2002	Sierra Club and Louisiana Environmental Action Network Inc. v. Gregg A. Cooke	Earthjustice Attorney Client Trust Account	496,375.61	Clean Water Act
12/20/2002	Mineral County v. EPA	Western Environmental Law Center	35,000.00	Clean Water Act
12/17/2002	American Littoral Society v. EPA	Mid-Atlantic Environmental Law Center	75,000.00	Clean Water Act
11/13/2002	Defend the Bay Inc. v. Marcus	Natural Resources Defense Council	57,500.00	Clean Water Act
10/30/2002	Sierra Club and Group Against Smog and Pollution v. EPA	Earthjustice Legal Defense Fund	125,000.00	Clean Air Act
10/30/2002	American Trucking Associations v. EPA	Utility Air Resources Group Escrow Account	92,250.00	Clean Air Act
10/8/2002	Environmental Defense Fund v. EPA	Blackburn Carter, PC	51,630.76	Clean Air Act

Date sent	Case title	Payee(s)	Payment amount[a]	Statute
10/7/2002	Midwest Ozone Group v. EPA	Jackson & Kelley, PLLC	22,000.00	Clean Air Act
10/4/2002	Californians for Alternatives for Toxics et al. v. Environmental Protection Agency	Brian Gaffney	56,500.00	Endangered SpeciesAct

Source: GAO analysis of Treasury Department data.

[a] Table data have not been adjusted for inflation.

[b] IOLTA stands for "interest on lawyers' trust accounts." According to IOLTA.org, a lawyer who receives funds that belong to a client, such as retainers or advance payments, must place those funds in a trust account separate from the lawyer's own money; these are known as IOLTA accounts.

Table 7. EPA's Equal Access to Justice Act Payments, Fiscal Year 2006 through Fiscal Year 2010

Payment date	Case title	Payee(s)	Payment amount[a]	Statute[b]
5/27/2010	Richard J. Weisberg v. EPA	Richard J. Weisberg	$1,178	Clean Water Act
4/8/2010	Center for Biological Diversity v. EPA	Center for Biological Diversity	18,000	Clean Water Act
9/29/2010	United Farm Workers of America v. EPA	United Farm Workers of America	230,000	Federal Insecticide,Fungicide, and Rodenticide Act
9/28/2009	Northwest Environmental Advocates v. EPA	Northwest Environmental Advocates[c]	465,000	Clean Water Act
5/21/2009	Kentucky Waterways Alliance v. Leavitt	Kentucky Waterways Alliance	147,638	Clean Water Act

Table 7. (Continued)

Payment date	Case title	Payee(s)	Payment amount[a]	Statute[b]
11/19/2008	Sierra Club v. EPA	Sierra Club Environmental Law Program/Mid-Atlantic Environmental Law Center	169,000	Clean Water Act
7/18/2008	Klamath Riverkeeper v. EPA	Klamath Riverkeeper	65,459	Clean Water Act
12/11/2007	National Wildlife Federation v. EPA	National Wildlife Federation	27,273	Clean Water Act
10/15/2007	Friends of the Earth v. EPA	Friends of the Earth	99,660	Clean Water Act
10/12/2007	Ohio Valley Environmental Coalition Inc. v. Johnson	Ohio Valley Environmental Coalition Inc.	2,805	Clean Water Act
9/18/2006	Iowa Environmental Council et al. v. Steven Johnson et al.	Iowa Environmental Council	8,414	Clean Water Act
7/24/2006	West Harlem Environmental Action v. EPA	West Harlem Environmental Action and National Resources Defense Council	72,000	Federal Insecticide,Fungicide, and Rodenticide Act
4/17/2006	St. Johns Riverkeeper and Linda Young v. EPA	St. Johns Riverkeeper	19,801	Clean Water Act
12/14/2005	Minnesota Center for Environmental Advocacy v. EPA	Minnesota Center for Environmental Advocacy	45,000	Clean Water Act

Source: GAO analysis of EPA data.

[a] These data have not been adjusted for inflation.

[b] Statute indicates the statute related to the EPA action being challenged. In some instances, the Administrative Procedure Act, rather than the statute listed, authorizes the lawsuit.

[c] One payment, the 2009 payment to the Northwest Environmental Advocates, is associated with two cases.

APPENDIX III: DEPARTMENT OF THE TREASURY AND ENVIRONMENTAL PROTECTION AGENCY PAYMENTS

This appendix provides data on payments for attorney fees and court costs made by the Department of the Treasury for fiscal year 2003 through fiscal year 2010 and by the Environmental Protection Agency (EPA) for fiscal year 2006 through fiscal year 2010. Payments for attorney fees and expenses and court costs may be made to a plaintiff or directly to a plaintiff's attorney. In cases involving multiple plaintiffs, one or more plaintiffs or their attorneys may receive payment. The first plaintiff named in the case title does not necessarily receive the payment. Table 6 shows payments from the Judgment Fund.

In addition to payments from the Judgment Fund, EPA made payments under EAJA to successful plaintiffs. Table 7 shows payments made by EPA for fiscal year 2006 through fiscal year 2010.

End Notes

[1]Some of these laws specifically authorize suits against EPA, while the Administrative Procedure Act, which is the federal law generally governing how federal agencies may propose and establish regulations, authorizes judicial review of certain federal agency actions.

[2] Treasury's Judgment Fund is also used to pay any court judgments that fall under Judgment Fund authorities. Justice's costs to defend EPA in Superfund litigation is reimbursed by EPA pursuant to an annual agreement between the agencies.

[3] GAO, *DOJ's Civil Rights Division: Opportunities Exist to Improve Its Case Management System and Better Meet Its Reporting Needs*, GAO-09-938R (Washington, D.C.: Sept. 30, 2009).

[4] From fiscal year 1993 to fiscal year 1994, the Attorney General reported annually to Congress on the amount of attorney fees and other expenses awarded in certain judicial proceedings. See GAO, *Private Attorneys: Selected Attorneys' Fee Awards against Nine Federal Agencies in 1993 and 1994*, GAO/GGD-96-18 (Washington, D.C.: Oct. 31, 1995).

[5] Actions that may be challenged in court generally fall into several categories: rulemakings, permit decisions and other approvals, enforcement actions, and other actions. In a rulemaking, EPA publishes a proposed regulation for public review and comment and then issues a final regulation. Generally, challenges may be brought after EPA has issued its final rule. In a permit decision, EPA processes an application according to relevant procedures, which typically provide for a draft permit and opportunity for the applicant and interested public to comment before the agency's issuance or denial of a final permit. Generally, only final permit decisions, including the process by which a decision was made, may be challenged.

[6] These environmental laws typically also authorize suits against other federal agencies for violations. For example, a citizen could file a lawsuit against a federal agency for operating a hazardous waste facility without a Resource Conservation and Recovery Act permit.

[7] Administrative Procedure Act, 60 Stat. 237 (1946), codified as amended at 5 U.S.C. § 551 (2011).

[8] Not all EPA actions are subject to legal challenge. For example, the Safe Drinking Water Act specifically precludes suits challenging EPA's list of contaminant candidates, a list periodically issued by EPA enumerating contaminants that may require regulation under the act.

[9] Committee on Public Works, *A Legislative History of the Clean Air Amendments of 1970*, at 351 (1974) (remarks of Senator Muskie). The Clean Air Act's citizen suit provision was the model for similar provisions in other environmental statutes.

[10] Edward Lloyd, *Citizen Suits and Defenses against Them*, ALI-ABA Environmental Litigation (2008). See also Robert Glicksman, *The Value of Agency-Forcing Citizen Suits to Enforce Nondiscretionary Duties*, 10 Widener L. Rev. 353 (2004). For a critical view of citizen suits under the Clean Air Act, see Randy E. Brogdon and Mack McGuffey, *Recent Trends in CAA Citizen Suits: Managing Risk in the Serengeti*, 20-WTR Nat'l Resources & Env't., 17 (2006).

[11] Generally, the environmental statutes' citizen-suit provisions require a prospective plaintiff to first send EPA a formal notice of intent to sue. Conversely, neither these statutes' judicial review provisions nor the Administrative Procedure Act impose a notice requirement.

[12] In general, a party must first exhaust all available administrative appeals before initiating a judicial suit.

[13] For example, EPA's Environmental Appeals Board can decide disputes such as appeals from permit decisions, civil penalty decisions, and other administrative decisions.

[14] In general, before the Equal Access to Justice Act (EAJA) was enacted, under the principle of sovereign immunity, the federal government was not subject to many of these exceptions and therefore was not authorized to make payments to prevailing parties. According to Justice, however, the relevant provisions of many key environmental statutes are applicable to the federal government, and thus the government is subject to court awards to prevailing parties independent of EAJA. See, e.g., Clean Air Act, 42 U.S.C. §§ 7604(d), 7607(f) (2011); Clean Water Act, 33 U.S.C. §§ 1365(d), 1369(b)(3) (2011).

[15] H.R. Conf. Rep. 96-1434, at 20-27 (1980) (Conference committee report on Pub. L. No. 96-481, The Small Business Export Expansion Act of 1980, of which Title II is the Equal Access to Justice Act).

[16] According to some in Congress, parties who choose to litigate an issue against the federal government help refine and formulate public policy by ensuring the legitimacy and fairness of the law being contested. Where policy changes are required, some believe that the costs should be borne by the government.

[17] For example, courts recognize a common-law exception allowing award of attorney fees to prevailing parties in instances of bad faith.

[18] This provision does not limit the eligibility of prevailing plaintiffs, and the statute requires that the fees be "reasonable" rather than impose a specific cap on the hourly rate of attorney fees for payment purposes. In addition, any fees awarded under this section are subject to any limitations that would apply to analogous awards against private parties, which may be provided by an underlying statute.

[19] EAJA authorizes the award of these fees against the federal government in both civil court actions, excluding tort cases, such as personal injury suits, as well as certain agency adjudications.

[20] This section limits the prevailing plaintiff's eligibility to receive payment by defining an eligible party as, at the time the lawsuit is filed, either an individual with a net worth below $2 million; or a business owner or any partnership, corporation, association, local government, or organization with a net worth below $7 million and fewer than 500 employees. However, tax-exempt nonprofit organizations and certain agricultural marketing cooperatives are considered parties regardless of net worth.

[21] Currently, there are no statutory requirements in effect for agency or governmentwide reporting of payments made under EAJA for either administrative or judicial proceedings. According to officials from the Administrative Conference of the United States, the conference has begun to obtain and compile such information for fiscal year 2010, noting that there has been continued interest in Congress (including pending legislation) regarding data on payments under EAJA. Officials told us the conference has requested EAJA data from 50 government agency conference members, as well as a few additional agencies that had previously reported EAJA activity to the conference. The chairman plans to publish a report for fiscal year 2010 later in 2011.

[22] GAO/GGD-96-18.

[23] GAO, *Equal Access to Justice Act: Its Use in Selected Agencies*, GAO/HEHS-98-58R (Washington, D.C.: Jan. 14, 1998).

[24] 28 U.S.C. § 2412(d) (2011).

[25] EPA's "credible evidence" rule, 62 Fed. Reg. 8314 (Feb. 24, 1997), allows any credible evidence to be used in enforcement actions related to operating permits under Clean Air Act emissions standards. Trade associations representing various industry groups, including car manufacturers, lumber companies, steel producers, petroleum companies, and mining companies challenged the rule in federal court. Twenty-five petitions were filed in the D.C. Court of Appeals, which consolidated them. See *Clean Air Implementation Project v. Environmental Protection Agency*, 150 F.3d 1200 (D.C. Cir. 1998).

[26] Except where otherwise noted, all figures are in constant 2010 dollars.

[27] We excluded data for fiscal years 1995 through 1997 from our cost estimate per case because the data for those years include only the number of cases filed during those years, not the total number of cases worked on that started in previous years. Including those years would make costs appear artificially low compared with later years, which include both new and continuing cases filed from 1995 onward. We estimated attorneys' costs by using the Justice data and making a number of adjustments.

[28] If all else remains the same, inclusion of cost such as the U.S. Attorneys' Offices' costs would increase our estimated costs.

[29] Of the approximately 2,500 cases, about 25 percent remain open in Justice's databases. Generally, payment of attorney fees and costs is among the last actions before a case is closed, although in some cases payments may be made while the case is open. For example, for a complex case, some parties may settle and receive payment, whereas the case may continue with other parties for several years. Thus, for the open cases, there could be associated attorney fees and cost payments in the future. Among only those cases that are closed in the databases, approximately 11 percent have a reported payment associated with them.

[30] For the District of Columbia district court and circuit court, Justice uses the Laffey matrix as a guide, but for other parts of the country, Justice consults with local U.S. Attorneys' Offices

about reasonable rates for that area or reviews relevant court decisions on attorney fee awards in that jurisdiction, since there is no single template of average rates to pay for attorney fees across the rest of the nation. The Laffey matrix provides hourly rates for attorneys of varying experience levels for the District of Columbia.

End Notes for Appendix II

[1] See, for example, Clean Air Act, 42 U.S.C. §§ 7604(d), 7607(f) (2011); Clean Water Act, 33 U.S.C. §§ 1365(d), 1369(b)(3) (2011).

[2] According to some in Congress, parties who choose to litigate an issue against the federal government help refine and formulate public policy by ensuring the legitimacy and fairness of the law being contested. Where policy changes are required, some believe that the costs should be borne by the government.

[3] For example, courts recognize a common-law exception allowing award of attorney fees in instances of bad faith.

[4] Statutes under which the agency may be sued include, but are not limited to, the Clean Air Act; Clean Water Act; Comprehensive Environmental Response, Compensation, and Liability Act (better known as the Superfund law); Emergency Planning and Community Right-to-Know Act; Federal Insecticide, Fungicide, and Rodenticide Act and related provisions of the Federal Food, Drug, and Cosmetic Act; Resource Conservation and Recovery Act; Safe Drinking Water Act; and Toxic Substances Control Act. In addition, EPA may be sued to challenge the agency's compliance with section 7 of the Endangered Species Act.

[5] EAJA authorized the award of these fees against the federal government in civil court actions, excluding tort cases such as personal injury suits.

[6] The Judgment Fund is not only the source of funds for these payments under this section of EAJA, but it is also used for many other payments stemming from court cases. For example, it is used to pay any court judgments for money against EPA stemming from an employment discrimination case or a contract dispute.

[7] GAO, *Private Attorneys: Selected Attorneys' Fee Awards against Nine Federal Agencies in 1993 and 1994*, GAO/GGD-96-018 (Washington, D.C.: Oct. 31, 1995).

[8] GAO, *Equal Access to Justice Act: Its Use in Selected Agencies*, GAO/HEHS-98-58R (Washington, D.C.: Jan. 14, 1998).

[9] 28 U.S.C. § 2412(d).

In: Environmental Litigation and the EPA
Editor: Martin F. Oliver
ISBN: 978-1-63482-720-1

Chapter 3

ENVIRONMENTAL LITIGATION: IMPACT OF DEADLINE SUITS ON EPA'S RULEMAKING IS LIMITED*

United States Government Accountability Office

ABBREVIATIONS

APA	Administrative Procedures Act
CFR	Code of Federal Regulations
CRA	Congressional Review Act
DOJ	Department of Justice
EPA	Environmental Protection Agency
EPCRA	Emergency Planning and Community Right-to-Know Act
MACT	Maximum Achievable Control Technology
NAAQS	National Ambient Air Quality Standards
NESHAP	National Emission Standards for Hazardous Air Pollutants
NO_2	Nitrogen Dioxide
NOI	Notice of Intent to Sue
NSPS	New Source Performance Standards
OAQPS	Office of Air Quality Planning and Standards
OAR	Office of Air and Radiation

* This is an edited, reformatted and augmented version of The United States Government Accountability Office publication, GAO-15-34, dated December 2014.

RTR	risk and technology review
SO_2	Sulfur Dioxide
VOC	Volatile Organic Compounds

WHY GAO DID THIS STUDY

Laws, such as the Clean Air Act, require EPA to issue rules by specific deadlines. Citizens can sue EPA for not issuing rules on time. These lawsuits are sometimes known as deadline suits. EPA sometimes negotiates a settlement to issue a rule by an agreed upon deadline. Some have expressed concern that the public is not involved in the negotiations and that settlements affect EPA rulemaking priorities. GAO was asked to review EPA settlements in deadline suits. This report examines (1) key environmental laws that allow deadline suits and the factors EPA and DOJ consider in determining whether to settle these suits, (2) the terms of settlements that led EPA to issue major rules in the last 5 years and the extent to which the public commented on the settlements, and (3) the extent to which settlements in deadline suits have affected EPA's rulemaking priorities.

GAO identified key laws allowing deadline suits through legal research and interviewed agency officials to understand the factors considered in determining whether to settle these suits. EPA identified the major rules it issued following settlements and GAO examined the text of those settlements. GAO examined EPA documentation to determine the extent to which the public commented on the settlements. Through data from EPA's Office of General Counsel and discussions with officials, GAO determined the extent to which settlements affected EPA's rulemaking priorities.

WHAT GAO RECOMMENDS

GAO is not making any recommendations in this report. DOJ and EPA concur with GAO's findings.

WHAT GAO FOUND

GAO identified seven key environmental laws that allow citizens to file a deadline suit against the Environmental Protection Agency (EPA) (see table) and EPA and the Department of Justice (DOJ) consider several factors in determining whether or not to settle these suits. The seven key environmental laws include, among others, the Clean Air Act and the Clean Water Act. EPA works with DOJ—which represents EPA in litigation —to decide whether to settle a deadline suit. EPA and DOJ officials stated that the factors they consider include (1) the cost of litigation, (2) the likelihood that EPA will win the case if it goes to trial, and (3) whether EPA and DOJ believe they can negotiate a settlement that will provide EPA with sufficient time to complete a final rule if required to do so.

Of the total number of major rules EPA promulgated from May 31, 2008 to June 1, 2013, nine were issued following seven settlements in deadline lawsuits, all under the Clean Air Act. The terms of the settlements in these deadline suits established a schedule to issue a statutorily required rule(s) or to issue a rule(s) unless EPA determined that doing so was not appropriate or necessary pursuant to the relevant statutory provision. None of the seven settlements included terms that finalized the substantive outcome of a rule. The Clean Air Act requires EPA to solicit public comments on drafts of settlements. The nine major rules were Clean Air Act rules, and EPA solicited public comments on all of the drafts. EPA received between 1 and 19 comments on six of the settlements and no comments on one settlement. EPA determined that none of the comments disclosed facts or other considerations compelling it to withdraw or withhold consent for the settlement.

The effect of settlements in deadline suits on EPA's rulemaking priorities is limited. According to EPA officials, settlements in deadline suits primarily affect a single office within EPA—the Office of Air Quality Planning and Standards (OAQPS)—because most deadline suits are based on provisions of the Clean Air Act for which that office is responsible. These provisions have recurring deadlines requiring EPA to set standards and to periodically review—and revise as necessary—those standards. OAQPS sets these standards through the rulemaking process. OAQPS officials said that deadline suits affect the timing and order in which rules are issued but not which rules are issued.

Key Environmental Laws That Allow Citizens to File Deadline Suits

Statute
Comprehensive Environmental Response, Compensation, and Liability Act (42 U.S.C.§§ 9601-9675)
Clean Air Act (42 U.S.C §§ 7401-7671q)
Clean Water Act (33 U.S.C. §§ 1251-1388)
Emergency Planning and Community Right-to-Know Act (42 U.S.C. §§ 11001-11050)
Safe Drinking Water Act(42 U.S.C. §§ 300f-300j-26)
Resource Conservation and Recovery Act (42 U.S.C. §§ 6901-6992k)
Toxic Substances Control Act (15 U.S.C. §§ 2601-2697)

Source: GAO. | GAO-15-34.

* * *

December 15, 2014

The Honorable Fred Upton
Chairman
Committee on Energy and Commerce
House of Representatives

The Honorable Ed Whitfield
Chairman
Subcommittee on Energy and Power Committee
on Energy and Commerce
House of Representatives

The Honorable Tim Murphy
Chairman
Subcommittee on Oversight and Investigations
Committee on Energy and Commerce
House of Representatives

Laws such as the Clean Air Act and the Clean Water Act require the Environmental Protection Agency (EPA) to take certain actions, such as issuing rules, to implement provisions of the law within certain statutorily designated time frames. In addition, some laws have provisions that allow any person or entity (including individuals, associations, businesses, and state and local governments) to file lawsuits to compel EPA to take statutorily required actions, such as issuing a rule, if it has not already done so within a statutorily designated time frame. Such lawsuits are sometimes referred to as deadline suits. When faced with deadline suits, EPA and the Department of Justice (DOJ)—which represents EPA in litigation—typically negotiate a settlement with the person or entity filing the lawsuit.

Some members of Congress and certain sectors of the public have expressed concern that, in settling these lawsuits, EPA is committing itself to significant new rulemakings and that affected stakeholders, and the public are unable to participate in the negotiating process before a settlement is finalized. They have also expressed concern about the impact of these settlements on EPA's rulemaking priorities.

Under the Congressional Review Act (CRA), all rules, including those issued by EPA, are classified as either a major rule or a nonmajor rule.

The CRA defines a "major" rule as one that has resulted in or is likely to result in (1) an annual effect on the economy of $100 million or more; (2) a major increase in costs or prices for consumers, individual industries, federal, state, or local government agencies, or geographic regions; or (3) significant adverse effects on competition, employment, investment, productivity, innovation, or on the ability of U.S.-based enterprises to compete with foreign-based enterprises in domestic and export markets.[1]

You asked us to review EPA's practices involving settlements in deadline suits that may compel EPA to issue a rule. This report examines (1) key environmental laws that allow citizens to file deadline suits that may compel EPA to conduct a rulemaking and the factors EPA and DOJ consider in determining whether or not to settle these lawsuits, (2) the terms of settlements in deadline suits that led EPA to promulgate major rules in the last 5 years and the extent to which the public commented on the terms of the settlements, and (3) the extent to which settlements in deadline suits have affected EPA's rulemaking priorities.

To address the first objective, we identified through legal research key environmental laws for which EPA has primary regulatory authority and the extent to which these laws allow citizens to file lawsuits that may compel agency rulemaking. We also interviewed EPA and DOJ officials to understand the factors EPA and DOJ consider in determining whether or not to settle deadline suits. To address the second objective, we accessed a database that GAO maintains to help implement the CRA to identify the major rules EPA issued from May 31, 2008 through June 1, 2013. We determined that the data were sufficiently reliable for the purpose of identifying major rules issued by EPA. EPA officials identified which of these rules EPA issued following a settlement in a deadline suit, and we obtained those settlements. We relied on EPA because neither EPA nor DOJ maintains a database that links settlements to rules, and there is no comprehensive public source of such information. We determined the extent to which the public commented on the terms of the settlements by examining EPA documentation on the number and content of the public comments. To address the third objective, we examined the extent to which settlements have affected EPA rulemaking priorities by obtaining data from EPA's Office of General Counsel on deadline suits EPA settled from January 2001 through July 2014 and the EPA offices responsible for implementing the terms of the settlements. We determined that the data were sufficiently reliable for the purpose of this report. The data showed that one EPA office was responsible for promulgating most of rules following settlements in deadline suits. We spoke with officials from this office to determine the extent to which settlements in deadline suits affected the office's rulemaking priorities. We also spoke with EPA budget officials knowledgeable about how EPA allocates resources among its offices to determine the extent to which deadline suits affect the resources allocated to offices not subject to deadline suits. Appendix I discusses our objectives, scope, and methodology in greater detail.

We conducted this performance audit from September 2013 to December 2014 in accordance with generally accepted government auditing standards. Those standards require that we plan and perform the audit to obtain sufficient, appropriate evidence to provide a reasonable basis for our findings and conclusions based on our audit objectives. We believe that the evidence obtained provides a reasonable basis for our findings and conclusions based on our audit objectives.

BACKGROUND

For the purposes of this report, we use the term deadline suit to mean a lawsuit in which an individual or entity sues because EPA has allegedly failed to perform any nondiscretionary act or duty by a deadline established in law. A nondiscretionary act or duty is an act or duty required by law. This report examines deadline suits that seek to compel EPA to either (1) issue a statutorily required rule when that rule has a deadline in law or (2) issue a statutorily required rule or make a determination that issuing such a rule is not appropriate or necessary pursuant to the relevant statutory provision, when issuing that rule or making that determination has a deadline in law. For example, a deadline suit may involve a person suing EPA because EPA failed to issue a rule by a date established in statute. Similarly, a person may sue EPA because it missed a recurring deadline to review and revise, as necessary or appropriate, an existing rule.[2] In August 2011, we reported that the number of new environmental litigation cases—of all types—filed against EPA each year from fiscal year 1995 through fiscal year 2010 averaged 155 cases per year.[3]

Before filing a deadline suit, a person generally must file a Notice of Intent to Sue (NOI) with EPA.[4] Among other things, a NOI generally must identify the provision(s) of the law that requires EPA to perform an act or duty and a description of the action taken or not taken by EPA that is claimed to constitute a failure to comply with the provision.[5] Sixty days after filing the NOI, the filer may initiate a deadline suit seeking a court order requiring EPA to complete the statutorily required action.[6]

A settlement takes the form of either a settlement agreement or a consent decree. For purposes of this report, the term settlement refers to both settlement agreements and consent decrees. Both are negotiated agreements between EPA and the plaintiff. A settlement agreement is not subject to court approval but can result in a stay of the lawsuit. If EPA fails to meet the terms of the settlement agreement, then the plaintiff can ask the court to lift the stay in order to proceed with the lawsuit. A consent decree is entered as a court order. If EPA fails to meet the terms of a consent decree, the court can enforce or modify the consent decree, including citing EPA for contempt of court.

Unless a more specific statute governs, when EPA or any other federal agency promulgates a rule, whether or not in conjunction with a deadline suit, it generally follows procedures prescribed in the Administrative Procedures Act (APA).[7] Among other things, the APA governs the process by which federal agencies develop and issue regulations. It includes requirements for publishing notices of proposed and final rules in the *Federal Register* and for providing opportunities for the public to comment on notices of proposed rulemaking. Many rules promulgated under the authority of the Clean Air Act do not follow the procedures prescribed in the APA, but rather follow similar but more specific procedures set forth in the act.[8]

KEY ENVIRONMENTAL LAWS ALLOW DEADLINE SUITS AND EPA CONSIDERS SEVERAL FACTORS IN RESPONDING TO THOSE SUITS

GAO identified seven key environmental laws that allow individuals to file a deadline suit to compel EPA to issue a statutorily required rule, or perform a statutorily required review of a rule to determine whether to revise the rule. EPA works with DOJ to consider several factors in determining whether or not to settle the deadline suit and the terms of any settlement.

Key Environmental Laws Allow Citizens to File Deadline Suits to Compel EPA Rulemaking

Table 1. Key Environmental Laws That Allow Citizens to File Deadline Suits

Statute	Citizen suit provision	General topic of the law
Comprehensive Environmental Response, Compensation, and Liability Act (42 U.S.C. §§ 9601-9675)	42 U.S.C. § 9659	Cleanup of hazardous substance releases
Clean Air Act (42 U.S.C. §§ 7401-7671q)	42 U.S.C. § 7604	Air quality
Clean Water Act (33 U.S.C. §§ 1251-1388)	33 U.S.C. § 1365	Water quality, including dredging, and filling of waters
Emergency Planning and Community Right-to-Know Act (42 U.S.C. §§ 11001-11050)	42 U.S.C. § 11046	Reporting releases of toxic chemicals
Safe Drinking Water Act (42 U.S.C. §§ 300f-300j-26)	42 U.S.C. § 300j-8	Water quality in public drinking water systems
Resource Conservation and Recovery Act (42 U.S.C. §§ 6901-6992k)	42 U.S.C. § 6972	Solid and hazardous waste storage, handling, treatment, and disposal
Toxic Substances Control Act (15 U.S.C. §§ 2601-2697)	15 U.S.C. § 2619	Toxic substances used in commerce

Source: GAO. | GAO-15-34.

GAO identified seven key environmental laws for which EPA has primary regulatory authority that allow citizens to file a deadline suit. Table 1 lists the seven laws.

With the exception of the Emergency Planning and Community Right-to-Know Act (EPCRA), the key environmental laws allow citizens to file deadline suits to compel EPA to perform any act or duty required by the respective law, including issuing any required rules. For example, the provision in the Clean Air Act states: "[A]ny person may commence a civil action on his own behalf - ... against the Administrator where there is alleged a failure of the Administrator to perform any act or duty under this chapter which is not discretionary with the Administrator."[9] The provision in EPCRA that allows citizens to file deadline suits is different from the other key environmental laws because citizen suits may only be filed to compel certain actions listed in the law.[10]

EPA and DOJ Consider Several Factors in Deciding Whether to Settle Deadline Suits

Within EPA, the Office of General Counsel is responsible for handling deadline suits. It works with the appropriate program offices in EPA, such as the Office of Air and Radiation (OAR) or the Office of Water, when negotiating settlements for deadline suits. EPA's Office of General Counsel also coordinates with DOJ's Environmental and Natural Resources Division.

According to EPA and DOJ officials, when a deadline suit is filed, the agencies work together to determine how to respond to the lawsuit, including whether or not to negotiate a settlement with the plaintiff or allow the lawsuit to proceed. In making this decision, EPA and DOJ consider several factors to determine which course of action is in the best interest of the government. According to EPA and DOJ officials, these factors include: (1) the cost of litigation, (2) the likelihood that EPA will win the case if it goes to trial, and (3) whether EPA and DOJ believe they can negotiate a settlement that will provide EPA with sufficient time to complete a final rule if required to do so. EPA and DOJ officials told us that they often choose to settle deadline suits when EPA has failed to fulfill a mandatory duty because it is very unlikely that the government will win the lawsuit. In many such cases, the only dispute is over the appropriate remedy, i.e., fixing a new date by which EPA should act. Additionally, in such cases, officials may believe that negotiating a settlement is the course of action most likely to create sufficient time for EPA to complete the rulemaking if it is required to issue a rule. EPA and DOJ have an agreement under which both must concur in the settlement of any case in which DOJ represents EPA.[11]

In negotiating the terms of settlements, EPA and DOJ are guided by, among other things, a 1986 DOJ memorandum—referred to as the Meese memorandum—the underlying concepts of which were codified in the Code of Federal Regulations (CFR) in 1991. [12] The regulation provides that any proposed settlement must be approved by the Deputy Attorney General or Associate Attorney General, as appropriate, when the proposed settlement converts an otherwise discretionary authority of the agency to promulgate, revise, or rescind regulations into a mandatory duty.[13] Thus, in general, this policy restricts DOJ from entering into a settlement if it commits EPA to take an otherwise discretionary action, such as including specific substantive content in the final rule unless an exception to this restriction is granted by the Deputy Attorney General or Associate Attorney General of the United States. According to EPA and DOJ officials, to their knowledge, EPA has been granted only one exception to the general restriction on creating mandatory duties through settlements—a 2008 settlement in a suit related to water quality criteria for pathogens and pathogen indicators.[14] The Meese memo also provides that DOJ should not enter into a settlement agreement that interferes with the agency's authority to revise, amend, or promulgate regulations through the procedures set forth in the APA.[15] As such, EPA officials stated that they have not, and would not agree to settlements in a deadline suit that finalizes the substantive outcome of the rulemaking or declare the substance of the final rule.

Settlements in Deadline Suits Established a Schedule for Issuing Rules

The terms of settlements in deadline suits that resulted in EPA issuing major rules in the last 5 years established a schedule to either promulgate a statutorily required rule or to promulgate a statutorily required rule or make a determination that doing so is not appropriate or necessary pursuant to the relevant statutory provision. EPA received public comments on all but one of the draft settlements in these suits.

SETTLEMENTS RESULTING IN MAJOR RULES INCLUDED RULEMAKING SCHEDULES, AND EPA RECEIVED PUBLIC COMMENTS ON DRAFTS OF MOST SETTLEMENTS

EPA issued 32 major rules from May 31, 2008 through June 1, 2013 (see app. II). According to EPA officials, the agency issued 9 of these rules following settlements in deadline suits. They were all Clean Air Act rules. The 9 rules stem from seven settlements. Two of the settlements established a schedule to complete 1 or more rules, and five established a schedule to complete 1 or more rules or make a determination that such a rule was not appropriate or necessary in accordance with the relevant statute. Some of the schedules included interim deadlines for conducting rulemaking tasks, such as publishing a notice of proposed rulemaking in the *Federal Register.*[16] Appendix III provides information on the schedules contained in each settlement.

In addition to schedules, the seven settlements also included, among other things, provisions that allowed deadlines to be modified (including the deadline to issue the final rule) and specified that nothing in the settlement can be construed to limit or modify any discretion accorded EPA by the Clean Air Act or by general principles of administrative law. Consistent with DOJ's 1986 Meese memorandum, none of the settlements we reviewed included terms that required EPA to take an otherwise discretionary action or prescribed a specific substantive outcome of the final rule.

The seven settlements, committing EPA to issue the 9 statutorily required rules, were finalized between about 10 months and more than 23 years after the applicable statutory deadlines. For each of the 9 rules, figure 1 shows the date the regulation was due, the date the settlement was filed with the court, and the date the final rule was published in the *Federal Register*.

EPA Received Public Comments on Drafts of Most Settlements in Deadline Suits

The Clean Air Act requires EPA, at least 30 days before a settlement under the act is final or filed with the court, to publish a notice in the *Federal Register* intended to afford persons not named as parties or interveners to the matter or action a reasonable opportunity to comment in writing.[17]

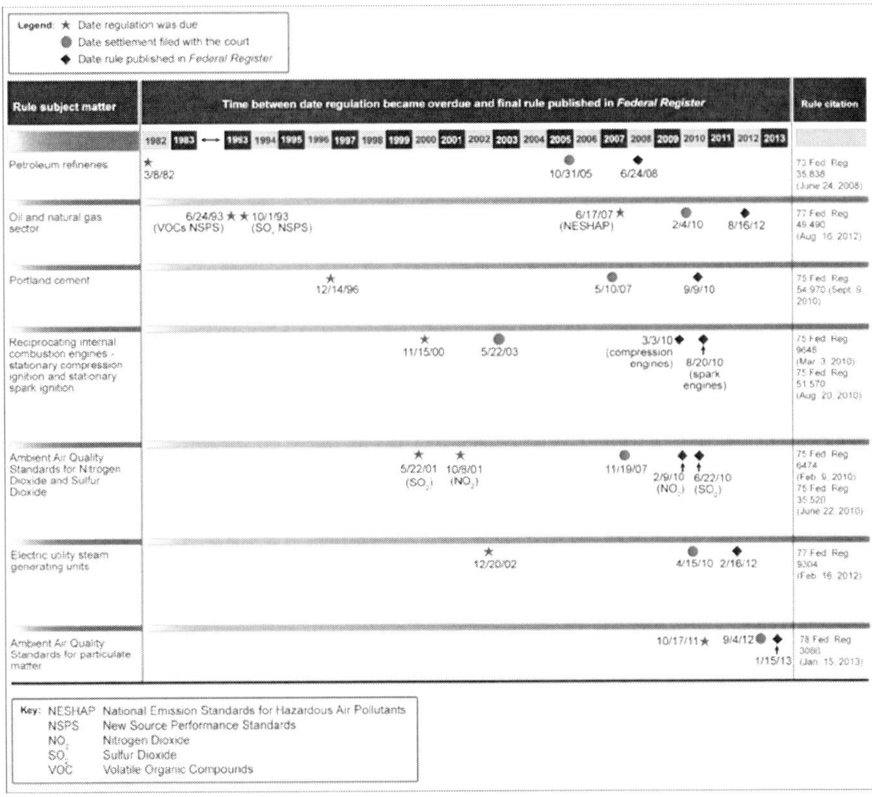

Source: GAO. | GAO-15-34.

Figure 1. Time Line of Key Dates Associated with 9 Major Clean Air Act Rules EPA Issued from May 31, 2008 to June 1, 2013.

EPA or DOJ, as appropriate, must then review the comments and may withdraw or withhold consent to the proposed settlement if the comments disclose facts or considerations that indicate consent to the settlement is inappropriate, improper, inadequate, or inconsistent with Clean Air Act requirements.[18] The other six key environmental laws with provisions that allow citizens to file deadline suits do not have a notice and comment requirement for proposed settlements.[19] According to an EPA official, with the exception of the agency's pesticide program, EPA generally does not ask for public comments on defensive settlements if the agency is not required to do so by statute.[20]

The 9 major rules EPA issued from May 31, 2008 to June 1, 2013 following seven settlements in deadline suits were Clean Air Act rules. For each settlement, EPA published a notice in the *Federal Register* providing the public the opportunity to comment on a draft of the settlement. EPA received between one and 19 public comments on six of the draft settlements. No comments were received on one of the draft settlements. Based on EPA summaries of the comments, the comments concerned the reasonableness of the deadlines contained in the settlements or supported or objected to the settlements. For example, some comments supported the deadline or asserted that the deadlines should be accelerated, others comments stated that EPA would have difficulty meeting the deadlines. EPA determined that none of the comments on any of the draft settlements disclosed facts or considerations that indicated that consent to the settlement in question would be inappropriate, improper, inadequate, or inconsistent with the act. Table 2 shows the number of public comments EPA received on each draft settlement.

Table 2. The Number of Public Comments EPA Received on Draft Settlements That Resulted in EPA Issuing 9 Clean Air Act Major Rules from May 31, 2008 to June 1, 2013

Rules issued following settlements in deadline suits	Number of public comments received on draft settlements
Standards of Performance for Petroleum Refineries, 73 Fed. Reg. 35,838 (June 24, 2008)	1
Primary National Ambient Air Quality Standards for Nitrogen Dioxide, 75 Fed. Reg. 6474 (Feb. 9, 2010) Primary National Ambient Air Quality Standard for Sulfur Dioxide, 75 Fed. Reg. 35,520 (June 22, 2010)	0
National Emission Standards for Hazardous Air Pollutants for Reciprocating Internal Combustion Engines, 75 Fed. Reg. 9648 (Mar. 3, 2010) National Emission Standards for Hazardous air Pollutants for Reciprocating Internal Combustion Engines, 75 Fed. Reg. 51,570 (Aug. 20, 2010)	5
National Emission Standards for Hazardous Air Pollutants from the Portland Cement Manufacturing Industry and Standards of Performance for Portland Cement Plants, 75 Fed. Reg. 54,970 (Sept. 9, 2010)	1 group provided 2 comments

Table 2. (Continued)

Rules issued following settlements in deadline suits	Number of public comments received on draft settlements
National Emission Standards for Hazardous Air Pollutants from Coal- and Oil-Fired Electric Utility Steam Generating Units and Standards of Performance for Fossil-Fuel-Fired Electric Utility, Industrial-Commercial-Institutional, and Small Industrial-Commercial-Institutional Steam Generating Units, 77 Fed. Reg. 9304 (Feb. 16, 2012)	9
Oil and Natural Gas Sector: New Source Performance Standards and National Emission Standards for Hazardous Air Pollutants Reviews, 77 Fed. Reg. 49,490 (Aug. 16, 2012)	1
National Ambient Air Quality Standards for Particulate Matter, 78 Fed. Reg. 3086 (Jan. 15, 2013)	19

Source: GAO. | GAO-15-34.

According to EPA Officials, Settlements in Deadline Suits Primarily Affect Rulemaking Priorities in a Single EPA Office

According to EPA officials, settlements in deadline suits primarily affect a single office within EPA—the Office of Air Quality Planning and Standards (OAQPS)—because most deadline suits are based on provisions of the Clean Air Act for which that office is responsible.[21] According to EPA's Office of General Counsel, provisions in the Clean Air Act that authorize the National Ambient Air Quality Standards (NAAQS) program and Air Toxics program account for most deadline suits. These provisions have recurring deadlines requiring EPA to set standards and to periodically review—and revise as appropriate or necessary—those standards. OAQPS sets these standards through the rulemaking process. For example, the Clean Air Act requires EPA to review and revise as appropriate NAAQS standards every 5 years and to review and revise as necessary technology standards for numerous air toxics generally every 8 years.

The effect of settlements in deadline suits on EPA's rulemaking priorities is limited. OAQPS officials said that deadline suits impact the timing and order in which rules are issued by the NAAQS program and the Air Toxics program, but not which rules are issued. The officials also noted that the impact of deadline suits on the two programs differs because of the different characteristics of the programs.

Regarding the NAAQS program, the Clean Air Act requires EPA to review and revise as appropriate the NAAQS standards for six pollutants—called criteria pollutants—at 5-year intervals.[22] NAAQS standards limit the allowable concentrations of criteria pollutants in the ambient air. There is more than one standard for each criteria pollutant.[23] EPA establishes the required standards through the rulemaking process and recently conducted seven NAAQS reviews to review the standards and revise as appropriate. According to an OAQPS official, prior to 2003, EPA did not review NAAQS on a regular cycle. Beginning in 2003, EPA faced four deadline suits for failure to complete NAAQS reviews for the six criteria pollutants. EPA settled two of these suits and was subject to a court order regarding the other two suits after it failed to successfully negotiate settlements with the plaintiffs. The settlements and court orders led EPA to perform the statutorily required reviews of the NAAQS standards for the six criteria pollutants and to promulgate seven rules—one for each NAAQS review.[24] The last of these seven rules was promulgated in April 2012.[25] According to officials, the deadline suits addressing the NAAQS standards did not affect which NAAQS standards were reviewed since EPA reviewed all of the standards. According to officials, the deadline suits did affect the timing and order in which EPA conducted the reviews to accommodate the time frames in the settlements and court orders. Additionally, according to officials, as a result of the experience in responding to the deadline suits, the agency is striving to maintain the 5-year statutory review cycle for criteria pollutants going forward. However, officials noted that it is difficult for EPA to complete its NAAQS reviews every 5 years. From April 2012 through September 2014, EPA has promulgated one rule following a NAAQS review after it settled a deadline suit and has missed the statutory deadline for reviewing the standards of two other criteria pollutants, one of which EPA is under court order to complete by October 2015 following a deadline suit.[26]

Regarding the Air Toxics program, OAQPS officials said that the impact of deadline suits on the Air Toxics program is different from that of the impact on the NAAQS program because of the large number of rules that the Air Toxics program promulgates. For example, the Clean Air Act establishes a schedule under which EPA established 120 standards to reduce the emissions of 187 hazardous air pollutants.[27] These National Emission Standards for Hazardous Air Pollutants (NESHAP) apply to certain categories of sources of these pollutants, such as cement manufacturing, municipal solid waste landfills, and semiconductor manufacturing. Generally, the act requires EPA, no less often than every 8 years, to review the standard, and revise as necessary. It makes any necessary revisions through the rulemaking process. The review must take into account developments in practices, processes, and control technologies.[28] For sources subject to Maximum Achievable Control Technology (MACT) standards promulgated pursuant to section 112(d)(2) of the Clean Air Act, EPA must also conduct a residual risk assessment within 8 years after the initial promulgation of the standard.[29] EPA refers to these two reviews together as the risk and technology review (RTR). As of October 2014, EPA has completed 28 RTRs (27 of these reviews following deadline suits) and has not completed 57 RTRs for which the statutory deadline has passed and 36 RTRs for which the statutory deadline has not passed.[30] Additionally, officials report that, currently, most of the resources available to complete RTRs are focused on a 2011 settlement. This settlement listed 27 NESHAPs for which RTRs were overdue. OAQPS officials said that they have been unable to meet all of the time frames contained in the 2011 settlement and, as a result, have negotiated amendments to the settlement extending the time frames. Officials said that they intend to complete all of the overdue RTRs but are focused on fulfilling the terms of the 2011 settlement and several other settlements to which EPA has entered into that address a smaller number of reviews. Additionally, in September 2013, EPA received a NOI concerning 43 additional NESHAPs for which a RTR is overdue. EPA officials said that they are engaged in settlement discussions over one of these reviews for which EPA has been sued.

Additionally, we discussed with EPA budget officials the potential impact of budget allocation decisions associated with deadline suits on EPA offices that are not subject to deadline suits. According to the budget officials, EPA accounts for anticipated workload arising out of litigation in its budgeting cycle for affected programs but does not make changes in existing budget allocations specifically to address settlements in deadline suits. Thus, according to the official, the resources available to EPA offices not subject to these settlements are not directly impacted by the settlements.

Agency Comments

We provided a draft of this report to EPA and DOJ for review and comment. In written comments from EPA, reproduced in appendix IV, the agency generally concurs with our analysis and states that the report accurately describes EPA's approach to deadline suit litigation brought against it. EPA also provided technical comments, which we incorporated as appropriate. In addition, in an e-mail received November 24, 2014, the DOJ Audit Liaison stated that the DOJ concurs with our report and has no additional comments.

J. Alfredo Gómez
Director, Natural Resources and Environment

Appendix I: Objectives, Scope, and Methodology

The objectives of this report are to examine (1) key environmental laws that allow citizens to file deadline suits that may compel the Environmental Protection Agency (EPA) to conduct a rulemaking and the factors EPA and the Department of Justice (DOJ) consider in determining whether or not to settle these lawsuits, (2) the terms of settlements in deadline suits that led EPA to promulgate major rules in the last 5 years and the extent to which the public commented on the terms of the settlements, and (3) the extent to which settlements in deadline suits have affected EPA's rulemaking priorities.

To examine the key environmental laws that allow citizens to file deadline suits that may compel EPA to conduct a rulemaking, we identified through legal research nine key environmental laws for which EPA has primary regulatory authority. Through additional legal research, we determined that two of these laws do not include provisions that permit citizens to file deadline suits. These laws are the Federal Insecticide, Fungicide, and Rodenticide Act and related provisions of the Federal Food, Drug, and Cosmetic Act.[1] To understand the factors that EPA considers in determining whether or not to settle deadline suits, we held discussions with officials from EPA's Office of General Counsel and DOJ because both agencies are involved in making these determinations. We also discussed the processes and procedures that EPA follows when settling citizen deadline suits.

To examine the terms of settlements in deadline suits that led EPA to promulgate major rules in the last 5 years, we developed a list of major rules

EPA issued from May 31, 2008 through June 1, 2013 by searching a database that GAO maintains to help implement the Congressional Review Act. [2] We determined that the data were sufficiently reliable for the purpose of identifying major rules issued by EPA. EPA officials then identified which major rules EPA issued following a settlement in a deadline suit. We relied on EPA because neither EPA nor DOJ maintains a database that links settlements to rules, and there is no comprehensive public source of such information. For the purposes of this report, we use the term deadline suit to mean a lawsuit in which an individual or entity sues because EPA has allegedly failed to perform any nondiscretionary act or duty by a deadline established in law. A nondiscretionary act or duty is an act or duty required by law. This report only examines deadline suits that seek to compel EPA to either (1) issue a statutorily required rule when that rule has a deadline in law or (2) issue a statutorily required rule or make a determination that issuing such a rule is not appropriate or necessary pursuant to the relevant statutory provision, when issuing that rule or making that determination has a deadline in law. We did not review other types of suits against EPA.[3] We obtained the settlements by accessing court records through the Public Access to Court Electronic Records. We then analyzed the content of each settlement and summarized the results.

To examine the extent to which the public commented on the terms of the settlements, we obtained from EPA legal memoranda summarizing the number and content of public comments EPA received on drafts of the settlements. Because each of the major rules issued following settlements in deadline suits were Clean Air Act rules, EPA solicited public comments on drafts of the settlements as required by the Clean Air Act. The act also requires EPA or DOJ to consider any public comments provided on settlements and authorizes them to withdraw or withhold consent to the proposed settlement if the comments disclose facts or considerations that indicate consent to the settlement is inappropriate, improper, inadequate, or inconsistent with the Clean Air Act requirements. EPA made these determinations and documented its decisions in legal memoranda that it provided to us. We analyzed the contents of these memoranda to determine the extent and nature of the public comments EPA received on draft settlements.

To examine the extent to which settlements in deadline suits have affected EPA's rulemaking priorities, we obtained from EPA's Office of General Counsel data on deadline suits it had settled from January 2001 through July 2014 and the EPA office(s) responsible for implementing the terms of the settlements.[4] We assessed the reliability of the data by interviewing agency officials knowledgeable about the data. We determined that the data were sufficiently reliable for the purposes of this report. The data showed that one office was responsible for implementing the terms of most of the settlements. We spoke with officials from this office to understand the extent to which settlements in deadline suits had affected the timing and order of the rules they promulgated, as well as which rules they promulgated. We also spoke with EPA budget officials to understand the extent to which settlements in deadline suits affected budget allocation decisions for EPA offices not subject to settlements in deadline suits.

We also interviewed individuals from academia, an environmental group, industry, and a state official from Oklahoma, to obtain their perspectives on deadline suits. We chose these individuals because they had experience or knowledge related to deadline suits and could provide the perspective of different stakeholder groups. For example, one interviewee provided legal representation for an environmental group that filed a deadline suit, and another interviewee authored a report critical of how EPA responds when faced with a deadline suits. The views of these individuals cannot be generalized to those with whom we did not speak.

We conducted this performance audit from September 2013 to December 2014 in accordance with generally accepted government auditing standards. Those standards require that we plan and perform the audit to obtain sufficient, appropriate evidence to provide a reasonable basis for our findings and conclusions based on our audit objectives. We believe that the evidence obtained provides a reasonable basis for our findings and conclusions based on our audit objectives.

APPENDIX II: MAJOR RULES ISSUED BY EPA FROM MAY 31, 2008 TO JUNE 1, 2013

The Environmental Protection Agency (EPA) issued 32 major rules from May 31, 2008 to June 1, 2013. According to EPA officials, the agency issued 9 of these rules following settlements in deadline suits and issued 5 of the 32 rules to comply with court orders following deadline suits in which plaintiffs and EPA were unable to reach a settlement. The remaining 18 rules, according to agency officials, were not associated with a deadline suit. Table 3 lists the 32 major rules EPA issued from May 31, 2008 to June 1, 2013.

Table 3. Major Rules Issued by EPA from May 31, 2008 to June 1, 2013

Rules issued following settlements in deadline suits	
Rule title	Law
Standards of Performance for Petroleum Refineries, 73 Fed. Reg. 35,838 (June 24, 2008)	Clean Air Act
Primary National Ambient Air Quality Standards for Nitrogen Dioxide, 75 Fed. Reg. 6474 (Feb. 9, 2010)	Clean Air Act
National Emission Standards for Hazardous Air Pollutants for Reciprocating Internal Combustion Engines, 75 Fed. Reg. 9648 (Mar. 3, 2010)	Clean Air Act
Primary National Ambient Air Quality Standard for Sulfur Dioxide, 75 Fed. Reg. 35,520 (June 22, 2010)	Clean Air Act
National Emission Standards for Hazardous Air Pollutants for Reciprocating Internal Combustion Engines, 75 Fed. Reg. 51,570 (Aug. 20, 2010)	Clean Air Act
National Emission Standards for Hazardous Air Pollutants From the Portland Cement Manufacturing Industry and Standards of Performance for Portland Cement Plants, 75 Fed. Reg. 54,970 (Sept. 9, 2010)	Clean Air Act
National Emission Standards for Hazardous Air Pollutants From Coal- and Oil-Fired Electric Utility Steam Generating Units and Standards of Performance for Fossil-Fuel-Fired Electric Utility, Industrial-Commercial-Institutional, and Small Industrial-Commercial-Institutional Steam Generating Units, 77 Fed. Reg. 9304 (Feb. 16, 2012)	Clean Air Act
Oil and Natural Gas Sector: New Source Performance Standards and National Emission Standards for Hazardous Air Pollutants Reviews, 77 Fed. Reg. 49,490 (Aug. 16, 2012)	Clean Air Act
National Ambient Air Quality Standards for Particulate Matter, 78 Fed. Reg. 3086 (Jan. 15, 2013)[a]	Clean Air Act
Rules issued following court orders in deadline suits	
National Ambient Air Quality Standards for Lead, 73 Fed. Reg. 66,964 (Nov. 12, 2008)	Clean Air Act
Effluent Limitations Guidelines and Standards for the Construction and Development Point Source Category, 74 Fed. Reg. 62,996 (Dec. 1, 2009)	Clean Water Act
National Emission Standards for Hazardous Air Pollutants for Area Sources: Industrial, Commercial, and Institutional Boilers, 76 Fed. Reg. 15,554 (Mar. 21, 2011)	Clean Air Act
National Emission Standards for Hazardous Air Pollutants for Major Sources: Industrial, Commercial, and Institutional Boilers and Process Heaters, 76 Fed. Reg. 15,608 (Mar. 21, 2011)	Clean Air Act
Standards of Performance for New Stationary Sources and Emission	Clean Air Act

Rules issued following settlements in deadline suits	
Rule title	Law
Guidelines for Existing Sources: Commercial and Industrial Solid Waste Incineration Units, 76 Fed. Reg. 15,704 (Mar. 21, 2011)	
Lead; Amendment to the Opt-out and Recordkeeping Provisions in the Renovation, Repair, and Painting Program, 75 Fed. Reg. 24,802 (May 6, 2010)	Toxic Substance Control Act
Standards of Performance for Petroleum Refineries; Standards of Performance for Petroleum Refineries for Which Construction, Reconstruction, or Modification Commenced After May 14, 2007, 77 Fed. Reg. 56,422 (Sept. 12, 2012)	Clean Air Act
National Emission Standards for Hazardous Air Pollutants for Reciprocating Internal Combustion Engines; New Source Performance Standards for Stationary Internal Combustion Engines 78 Fed. Reg. 6674 (Jan. 30, 2013)	Clean Air Act
Control of Emissions From Nonroad Spark-Ignition Engines and Equipment, 73 Fed. Reg. 59,034 (Oct. 8, 2008)	Clean Air Act
Revisions to the Definition of Solid Waste, 73 Fed. Reg. 64,668 (Oct. 30, 2008)	Resource Conservation and Recovery Act
Oil Pollution Prevention; Spill Prevention, Control, and Countermeasure Rule Requirements – Amendments, 73 Fed. Reg. 74,236 (Dec. 5, 2008)	Clean Water Act
Mandatory Reporting of Greenhouse Gases, 74 Fed. Reg. 56,260 (Oct. 30, 2009)	Clean Air Act
Oil Pollution Prevention; Spill Prevention, Control, and Countermeasure Rule Requirements – Amendments, 74 Fed. Reg. 58,784 (Nov. 13, 2009)	Clean Water Act
Regulation of Fuels and Fuel Additives: Changes to Renewable Fuel Standard Program, 75 Fed. Reg. 14,670 (Mar. 26, 2010)	Clean Air Act
Light-Duty Vehicle Greenhouse Gas Emission Standards and Corporate Average Fuel Economy Standards, 75 Fed. Reg. 25,324 (May 7, 2010)	Clean Air Act
Prevention of Significant Deterioration and Title V Greenhouse Gas Tailoring Rule, 75 Fed. Reg. 31,514 (June 3, 2010)	Clean Air Act
Oil Pollution Prevention: Spill Prevention, Control, and Countermeasure (SPCC) Rule - Amendments for Milk and Milk Product Containers, 76 Fed. Reg. 21,652 (Apr. 18, 2011)	Clean Water Act
Federal Implementation Plans: Interstate Transport of Fine Particulate Matter and Ozone and Correction of SIP Approvals, 76 Fed. Reg. 48,208 (Aug. 8, 2011)	Clean Air Act
Greenhouse Gas Emissions Standards and Fuel Efficiency Standards for Medium- and Heavy-Duty Engines and Vehicles, 76 Fed. Reg. 57,106 (Sept. 15, 2011)	Clean Air Act

Table 3. (Continued)

Rules issued following settlements in deadline suits	
Rule title	Law
Federal Implementation Plans for Iowa, Michigan, Missouri, Oklahoma, and Wisconsin and Determination for Kansas Regarding Interstate Transport of Ozone, 76 Fed. Reg. 80,760 (Dec. 27, 2011)	Clean Air Act
Regulation of Fuels and Fuel Additives: 2013 Biomass-Based Diesel Renewable Fuel Volume, 77 Fed. Reg. 59,458 (Sept. 27, 2012)	Clean Air Act
2017 and Later Model Year Light-Duty Vehicle Greenhouse Gas Emissions and Corporate Average Fuel Economy Standards, 77 Fed. Reg. 59,458 (Sept. 27, 2012)	Clean Air Act
National Emission Standards for Hazardous Air Pollutants for Major Sources: Industrial, Commercial, and Institutional Boilers and Process Heaters, 78 Fed. Reg. 7138 (Jan. 31, 2013)	Clean Air Act

Source: GAO. | GAO-15-34.

[a]This rule was issued after the court issued an order establishing a deadline to sign the notice of proposed rulemaking, and the parties reached a settlement establishing a deadline to sign the final rule.

APPENDIX III: CONTENTS OF SETTLEMENTS FOLLOWING DEADLINE SUITS THAT COMPELLED EPA TO ISSUE 9 RULES FROM MAY 31, 2008 TO JUNE 1, 2013

The Environmental Protection Agency (EPA) issued 9 major rules from May 31, 2008 to June 1, 2013 following seven settlements in deadline suits. Each of the seven settlements established a schedule to either issue a statutorily required rule or make a determination that such a rule is not appropriate or necessary pursuant to the relevant statutory provision. EPA negotiated extensions to the deadlines to issue the final rules in five of the seven settlements. Table 4 summarizes the contents of the seven settlements.

Table 4. Contents of Settlements in Deadline Suits Associated with 9 Major Rules EPA Issued from May 31, 2008 to June

Settlements containing a schedule to issue a rule	
Rules associated with the settlement	Settlement provisions and amendments
National Emission Standards for Hazardous Air Pollutants	Among other things, deadlines for (1) proposing and (2) promulgating a final rule for stationary internal combustion engines that are area

Settlements containing a schedule to issue a rule	
Rules associated with the settlement	Settlement provisions and amendments
for Reciprocating Internal Combustion Engines, 75 Fed. Reg. 9648 (Mar. 3, 2010) National Emission Standards for Hazardous air Pollutants for Reciprocating Internal Combustion Engines: Existing Stationary Spark Ignition, 75 Fed. Reg. 51,570 (Aug. 20, 2010)	sources or are less than 500 horsepower and located at major sources. The rule for stationary internal combustion engines that are area sources or less than 500 horsepower and located at major sources was subsequently divided into new sources and existing sources, and the deadline for issuing a final rule for existing sources was extended. The rule for existing sources was then divided into rules for compression-ignition engines and spark ignition engines, and the deadline for issuing each rule was extended, making the final deadlines 26 months and 32 months, respectively, later than the initial deadline in the settlement.
National Emission Standards for Hazardous Air Pollutants from Coal- and Oil-Fired Electric Utility Steam Generating Units and Standards of Performance for Fossil-Fuel-Fired Electric Utility, Industrial-Commercial-Institutional, and Small Industrial-Commercial-Institutional Steam Generating Units, 77 Fed. Reg. 9304 (Feb. 16, 2012)	Deadlines for signing for publication in the *Federal Register* a notice of (1) proposed and (2) final rulemaking setting forth EPA's emission standards for coal- and oil-fired electric utility steam generating units. The deadline for signing a notice of final rulemaking was subsequently extended by one month.
Settlements containing a schedule to issue a rule or make a determination that a rule is not necessary or appropriate	
Rules associated with the settlement	Settlement provisions and amendments
Standards of Performance for Petroleum Refineries, 73 Fed. Reg. 35,838 (June 24, 2008)	Deadlines for signing for publication in the *Federal Register*(1) proposed revisions to standards in New Source Performance Standards (NSPS), except to the extent that EPA determines that a review and/or revision is not appropriate, and (2) final revisions to standards in NSPS, except to the extent that EPA determines that a review and/or revision is not appropriate.
Primary National Ambient Air Quality Standards for Nitrogen Dioxide, 75 Fed. Reg. 6474 (Feb. 9, 2010) Primary National Ambient Air Quality Standard for Sulfur Dioxide, 75 Fed. Reg. 35,520 (June 22, 2010)	Deadlines for (1) issuing final Integrated Science Assessments addressing human health effects of oxides of nitrogen and sulfur, (2) signing a notice of proposed rulemaking concerning its review of primary oxides of nitrogen and sulfur National Ambient Air Quality Standards (NAAQS) and proposing such new and/or revisions of primary oxides of nitrogen and sulfur NAAQS as may be
	appropriate pursuant to the law, and (3) signing a notice of final rulemaking concerning its review of primary oxides of nitrogen and sulfur NAAQS and making such revisions and/or promulgating new primary oxides of nitrogen and sulfur NAAQS as may be appropriate pursuant to the law.[a]

Table 4. (Continued)

Settlements containing a schedule to issue a rule	
Rules associated with the settlement	Settlement provisions and amendments
	The deadlines for issuing proposed and final rules for primary NAAQS for oxides of nitrogen and sulfur were subsequently extended by between about 1 and 4 months.
National Emission Standards for Hazardous Air Pollutants from the Portland Cement Manufacturing Industry and Standards of Performance for Portland Cement Plants, 75 Fed. Reg. 54,970 (Sept. 9, 2010)	Deadlines for signing and forward to the Federal Registerfor publication (1) one or a combination of the following: a proposed rule containing revisions to NSPS and/or a proposed determination not to revise NSPS and (2) one or a combination of the following: a final rule containing revisions to the NSPS and/or a final determination not to revise NSPS. The deadline for issuing a final rule was subsequently extended by about 15 months.
Oil and Natural Gas Sector: New Source Performance Standards and National Emission Standards for Hazardous Air Pollutants Reviews, 77 Fed. Reg. 49,490 (Aug. 16, 2012)	Deadlines for (1) signing a proposed rule containing revisions to NSPS, a proposed determination not to revise the NSPS, and/or a proposed or final determination not to review the NSPS, (2) signing a final rule containing revisions to NSPS, a final determination not to revise NSPS, and/or a final determination not to review NSPS, (3) signing a proposed rule containing revisions to National Emission Standards for Hazardous Air Pollutants (NESHAP) or a proposed determination that the revision of NESHAP is not necessary, (4) signing a final rule containing revisions to NESHAP or a final determination that revision of NESHAP is not necessary, (5) signing a proposed rule promulgating residual risk standards or a proposed determination that promulgation of standards is not required, and (6) signing a final rule promulgating residual risk standards or a final determination that promulgation of standards is not required. The deadlines were subsequently extended by about 4 to 6 months.
National Ambient Air Quality Standards for Particulate Matter, 78 Fed. Reg. 3086 (Jan. 15, 2013)	Deadline for signing a notice of final rulemaking setting forth its final decision concerning its review of the NAAQS for particulate matter and promulgating such revisions to the NAAQS, and/or promulgating such new NAAQS as may be appropriate

Source: GAO. | GAO-15-34.

[a] Integrated Science Assessments evaluate and synthesize the most policy-relevant science for reviewing the National Ambient Air Quality Standards (NAAQS). The Clean Air Act requires EPA to review, and revise as appropriate, NAAQS at 5-year intervals. 42 U.S.C. § 7409(d).

[b] The court had previously issued an order establishing a deadline to sign the notice of proposed rulemaking.

APPENDIX IV: COMMENTS FROM THE ENVIRONMENTAL PROTECTION AGENCY

UNITED STATES ENVIRONMENTAL PROTECTION AGENCY
Washington, D.C. 20460

OFFICE OF
GENERAL COUNSEL

NOV 25 2014

Mr. J. Alfredo Gomez
Director, Natural Resources and Environment
U.S. Government Accountability Office
441 G Street, N.W.
Washington, DC 20460

Dear Mr. Gomez:

Thank you for the opportunity to review and comment on the U.S. Government Accountability Office's draft report, "Environmental Litigation: Impact of Deadline Suits on EPA's Rulemaking is Limited" (GAO-15-34). EPA generally concurs in your analysis.

This report describes our longstanding and entirely appropriate approach to litigation brought against the U.S. Environmental Protection Agency (EPA). We, alongside the U.S. Department of Justice (DOJ), carefully and completely assess the merits of any claims against EPA, our litigation risks based on the law and underlying facts, and the interests of the Agency and the American public. We never agree to final substantive outcomes or otherwise use settlements to circumvent the lawful process for promulgating environmental regulations and standards. Finally, all of the settlements that led to major rules in your study were subjected to public scrutiny and comment before being finalized.

While we are always interested in ways to improve transparency where appropriate, many opportunities for robust public participation in the work of EPA already exist. These include opportunities to comment on many proposed settlements before they are finalized, and the comment period provided in any subsequent rulemaking process. Indeed, the latter opportunity is used by millions of Americans, from every State and all walks of life, to help inform the Agency in developing its rules in accordance with the law and the underlying science.

It is my hope that this report will put to rest any misconceptions about the relationship between the manner in which EPA and DOJ litigate claims brought against EPA and the content of any EPA regulations. In recent years, this Agency and others within the federal government have faced unfounded allegations that our efforts to resolve litigation are somehow conducted in a manner to evade public input on important questions of public policy. As your findings confirm, nothing could be further from the truth.

The fact is, Congress (wisely, in my view) designed key aspects of our environmental laws to mandate regulatory action by certain deadlines to improve the quality of the environment and the protection of public health. There is nothing secret about those mandates; they are the law of the land. And it was Congress that gave the public the legal tools to hold the government to account in carrying out the mandates it set.

There should be no doubt of the diligence and commitment of the attorneys in my office and at the Department of Justice, to carry out our laws fully and faithfully. EPA strives to meet statutory deadlines. When clear deadlines are missed, and EPA is sued on that basis, the sound judgment of EPA and DOJ—as expressed from Administration to Administration—and the reaction of the courts, make clear that it is almost always a fruitless endeavor to seek to avoid those deadlines through legal maneuvering. Further, such efforts cost the taxpayers' money and squander our limited resources. Settlements can allow EPA to avoid unnecessary costs and litigation and focus its resources, and DOJ's, on more productive efforts to advance and defend the Agency's work.

I also want to take this opportunity to thank you and your staff for your professionalism, objectivity, and diligence. By zealously researching the facts and circumstances surrounding our deadline suit litigation, you have provided a valuable contribution to public understanding of how EPA meets its statutory mandates to protect human health and the environment.

If you have questions or need further information regarding this engagement, please contact Marna McDermott ((202)-564-2890).

Sincerely,

Avi S. Garbow
General Counsel

cc: Gina McCarthy, Administrator
Joel Beauvais, Associate Administrator, Office of Policy
Janet McCabe, Assistant Administrator, Office of Air and Radiation
Sam Hirsch, Acting Assistant Attorney General for Environment and Natural Resources
Lee J. Lofthus, Assistant Attorney General for Administration

2

End Notes

[1] 5 U.S.C. § 804(2). The CRA requires agencies to submit final rules, along with supporting information, to Congress and GAO for review before they can take effect. GAO's primary role under CRA is to provide Congress with a report on each major rule containing GAO's

assessment of whether the promulgating federal agency's submissions to GAO indicate that it has complied with the procedural steps required by various acts and executive orders governing the regulatory process. Congress then has an opportunity to review a major rule before it takes effect and to disapprove any rule to which it objects through the passage of a joint resolution with presentment to the President. 5 U.S.C. § 801.

[2] This report examines nondiscretionary duty suits in which the nondiscretionary duty is to issue a rule or make a determination not to issue a rule by a deadline in law. However, there are other types of nondiscretionary duties for which EPA can be sued. For example, in general, if a provision in law requires an agency to take an action, such as respond to a petition, but does not specify a date or set amount of time after the occurrence of an event by which the action must be taken, then a lawsuit to compel such an action is often referred to as an unreasonable delay suit. In addition to deadline and unreasonable delay suits, EPA can also be sued on other grounds. For example, a final rule issued by EPA may be challenged in court. As a result, a court may order EPA, or EPA may agree in a settlement, to issue a new rule. We did not include suits other than deadline suits in our review.

[3] GAO, Environmental Litigation: Cases Against EPA and Associated Costs over Time, GAO-11-650 (Washington, D.C.: Aug. 1, 2011).

[4] See, e.g., 42 U.S.C. § 7604(b)(2).

[5] See, e.g., 40 C.F.R. § 54.3(a). Although not statutorily required to publically post NOIs, in January 2013, EPA began to post NOIs at http://www.epa.gov/ogc/noi.html.

[6] See, e.g., 42 U.S.C. § 7604(a), (b)(2).

[7] 5 U.S.C. § 553.

[8] 42 U.S.C. § 7607(d).

[9] 42 U.S.C. § 7604(a)(2). Issuing a statutorily required rule is an example of a nondiscretionary act.

[10] See 42 U.S.C. § 11046(a)(1)(B).

[11] Memorandum of Understanding Between the Department of Justice and the Environmental Protection Agency (June 15, 1977).

[12] See 28 C.F.R. §§ 0.160-0.163.

[13] The Meese memorandum and the regulation treat the distinction between consent decrees and settlement agreements differently. The Meese memorandum draws a distinction between the executive branch's authority to enter into consent decrees, which are negotiated agreements that are formally entered as court orders and thus enforceable by the court, and settlement agreements, which are similar in form to consent decrees but not entered as a court order. According to the memorandum, DOJ should not enter into a consent decree that converts into a mandatory duty the otherwise discretionary authority of that agency to promulgate, revise, or rescind regulations. Regarding settlement agreements, the memorandum states that DOJ should not enter into a settlement agreement that interferes with the agency's authority to revise, amend, or promulgate regulations through the procedures set forth in the Administrative Procedures Act (APA). The related regulation does not make a distinction between consent decrees and settlement agreements. It requires that all proposed settlements that convert into a mandatory duty the otherwise discretionary authority of that agency to promulgate, revise, or rescind regulations be referred to the Deputy Attorney General or the Associate Attorney General of the United States. Finally, a 1999 DOJ memorandum—referred to as the Moss memo—concludes that the distinction that the Meese memorandum draws between consent decrees and settlement agreements is not of legal significance for purposes of determining the legal limits on discretion-limiting settlements except, possibly, in rare cases.

[14] See Natural Resources Defense Council et al. vs. Johnson et al., No. CV06-04843 PSG (JTL) (C.D.Cal. 2008). In this settlement, EPA agreed to complete two epidemiological studies not expressly required by section 104(v) of the Clean Water Act after a court ruled that EPA had not satisfied the requirements of that section.

[15] See 5 U.S.C. § 553.

[16] Even where interim rulemaking steps are not expressly set forth in a settlement, EPA officials stated that they go through the necessary rulemaking procedures such as notice and comment.

[17] See 42 U.S.C. § 7413(g).

[18] See id.

[19] The APA and EPA regulations require notice and comment of proposed agency rulemaking, whether or not it may have been the subject of a settlement. 5 U.S.C. § 553(b), (c); 40 C.F.R. § 25.10. This includes the right of interested persons to be given an opportunity to participate in the rulemaking by the submission of written data, views, and arguments. According to DOJ officials, settlements are structured to preserve public participation in any ensuing rulemaking proceeding and to not predetermine the outcome of that proceeding.

[20] According to the EPA official, the pesticides program has a regular practice of posting on its website proposed settlements associated with the program's issues. The program posts these documents because the program has a well-defined and knowledgeable community of stakeholders that is likely to be affected by the settlements.

[21] OAQPS is an office within EPA's Office of Air and Radiation.

[22] 42 U.S.C. § 7409(d). The six criteria pollutants are: carbon monoxide, lead, oxides of nitrogen, ozone, particulate matter, and oxides of sulfur.

[23] The Clean Air Act requires EPA to set primary standards and secondary standards for each criteria pollutant. Primary standards protect public health and secondary standards protect the public welfare. 42 U.S.C § 7409(a), (b).

[24] Because EPA decided to conduct a separate, coordinated review of the public welfare effects of oxides of sulfur and nitrogen rather than reviewing the public welfare effects of each at the same time it reviewed the respective public health effects, EPA promulgated 7 rules for the six criteria pollutants.

[25] 77 Fed. Reg. 20,218 (Apr. 3, 2012). This was not a major rule, as defined by the CRA.

[26] EPA promulgated a rule revising standards for particulate matter in January 2013 following a court order and consent decree and is under a separate court order to complete a review of ozone by October 2015. See 78 Fed. Reg. 3086 (Jan. 15, 2013); American Lung Assoc. and National Parks Conservation Assoc. v. EPA, No. 12-cv-00243 (D.D.C. 2012); Sierra Club v. EPA, No. 4:13-cv-02809-YGR (N.D. Cal. 2012). The deadline to review the standards for lead and, if needed, promulgate a rule revising the standards was November 2013. See 73 Fed. Reg. 66,964 (Nov. 12, 2008) (revising the primary and secondary NAAQS for lead).

[27] See 42 U.S.C. § 7412. For the purposes of this report, generally, we are counting each subpart of 40 C.F.R. part 63 as a NESHAP.

[28] 42 U.S.C. § 7412(d)(6).

[29] 42 U.S.C. § 7412(f)(2)(A) (requiring EPA, within 8 years after promulgation of standards for each category or subcategory of sources, to promulgate standards for such category or subcategory if promulgation of such standards is required in order to provide an ample margin of safety to protect public health or to prevent, taking into consideration costs, energy, safety, and other relevant factors, an adverse environmental effect). Standards that

apply only to area sources may not need to undergo a residual risk assessment, in which case these standards only receive a technology review every 8 years. See 42 U.S.C. § 7412(f)(5).

[30] These numbers do not add to the total number of standards because for one standard, an RTR was conducted for some source categories but not for others. As a result, that standard is counted both as having a completed RTR and having an overdue RTR.

End Notes for Appendix I

[1] 7 U.S.C. §§ 136-136y; 21 U.S.C. § 346a. The laws do not have provisions allowing citizens to sue EPA for failure to issue a statutorily required rule within statutorily required time frames, but the Administrative Procedures Act provides that a person suffering legal wrong because of agency action, or adversely affected or aggrieved by agency action within the meaning of a relevant statute, is entitled to judicial review thereof. 5 U.S.C. § 702. In such a case, the court may compel agency action unlawfully withheld or unreasonably delayed. 5 U.S.C. § 706.

[2] The act defines a "major" rule as one that has resulted in or is likely to result in (1) an annual effect on the economy of $100 million or more; (2) a major increase in costs or prices for consumers, individual industries, federal, state, or local government agencies, or geographic regions; or (3) significant adverse effects on competition, employment, investment, productivity, innovation, or on the ability of U.S.-based enterprises to compete with foreign-based enterprises in domestic and export markets. 5 U.S.C. § 804(2).

[3] This report examines nondiscretionary duty suits in which the nondiscretionary duty is to issue a rule or make a determination not to issue a rule by a deadline in law. However, there are other types of nondiscretionary duties for which EPA can be sued. For example, in general, if a provision in law requires an agency to take an action, such as respond to a petition, but does not specify a date or set amount of time after the occurrence of an event by which the action must be taken, then a lawsuit to compel such an action is often referred to as an unreasonable delay suit. In addition to deadline and unreasonable delay suits, EPA can also be sued on other grounds. For example, a final rule issued by EPA may be challenged in court. As a result, a court may order EPA, or EPA may agree in a settlement, to issue a new rule.

[4] We excluded from our analysis settlements in deadline suits for which a regional office was responsible for issuing the required rules because these rules are regional in scope.

INDEX

S

T

U

V

W